# Transactions on Computer Systems and Networks

**Series Editor**

Amlan Chakrabarti, Director and Professor, A. K. Choudhury School of Information Technology, Kolkota, West Bengal, India

Transactions on Computer Systems and Networks is a unique series that aims to capture advances in evolution of computer hardware and software systems and progress in computer networks. Computing Systems in present world span from miniature IoT nodes and embedded computing systems to large-scale cloud infrastructures, which necessitates developing systems architecture, storage infrastructure and process management to work at various scales. Present day networking technologies provide pervasive global coverage on a scale and enable multitude of transformative technologies. The new landscape of computing comprises of self-aware autonomous systems, which are built upon a software-hardware collaborative framework. These systems are designed to execute critical and non-critical tasks involving a variety of processing resources like multi-core CPUs, reconfigurable hardware, GPUs and TPUs which are managed through virtualisation, real-time process management and fault-tolerance. While AI, Machine Learning and Deep Learning tasks are predominantly increasing in the application space the computing system research aim towards efficient means of data processing, memory management, real-time task scheduling, scalable, secured and energy aware computing. The paradigm of computer networks also extends it support to this evolving application scenario through various advanced protocols, architectures and services. This series aims to present leading works on advances in theory, design, behaviour and applications in computing systems and networks. The Series accepts research monographs, introductory and advanced textbooks, professional books, reference works, and select conference proceedings.

More information about this series at https://link.springer.com/bookseries/16657

Amitava Choudhury · Arindam Biswas ·
T. P. Singh · Santanu Kumar Ghosh
Editors

# Smart Agriculture Automation Using Advanced Technologies

## Data Analytics and Machine Learning, Cloud Architecture, Automation and IoT

*Editors*
Amitava Choudhury
School of Computer Science
University of Petroleum and Energy Studies
Dehradun, India

Arindam Biswas
School of Mines and Metallurgy
Kazi Nazrul University
Asansol, West Bengal, India

T. P. Singh
School of Computer Science
University of Petroleum and Energy Studies
Dehradun, Uttarakhand, India

Santanu Kumar Ghosh
Department of Mathematics
Kazi Nazrul University
Asansol, India

ISSN 2730-7484 ISSN 2730-7492 (electronic)
Transactions on Computer Systems and Networks
ISBN 978-981-16-6123-5 ISBN 978-981-16-6124-2 (eBook)
https://doi.org/10.1007/978-981-16-6124-2



This Springer imprint is published by the registered company Springer Nature Singapore Pte Ltd.
The registered company address is: 152 Beach Road, #21-01/04 Gateway East, Singapore 189721, Singapore

# Preface

Agriculture automation is the primary concern and emerging subject for every country. The world population is increasing rapidly, and with the increase in population, the need for food rises briskly. Traditional methods used by farmers aren't sufficient to serve the increasing demand, so they have to hamper the soil by using harmful pesticides in an intensified manner. This affects the agricultural practice a lot, and in the end, the land remains barren with no fertility. This book talks about different automation practices like IoT, wireless communications, machine learning, artificial intelligence and deep learning. Some areas are causing problems in the agriculture field like crop diseases, lack of storage management, pesticide control, weed management, lack of irrigation and water management. All these problems can be solved by the techniques mentioned earlier.

Agriculture is the essential and most vital profession of our nation because it balances food necessity and the fundamental crude materials for a few enterprises and there is a consistent expansion sought after populace development. Various everyday issues are getting an incredible effect by the progressions in innovation. For a few years, individuals are running after the robotization with some degree of knowledge to supplant or limit humans from process cycles. Innovation in agriculture lessens reliance on individual human work and land. The invention permits operational concocting and speeds up decision-making on the farms. Ongoing headways in creation have an extraordinary effect on agriculture, and it has been set up that IoT is utilized in cultivating to improve the nature of farming. Advancement of Machine Learning (ML) and Internet of Things (IoT) has assembled consideration of specialists to apply these strategies in fields like farming. It causes ranchers to expand the profitability of their property so the overall demand for food can be satisfied. IoT is a high-level innovation for observing and controlling gadgets anyplace on the planet. It can interface devices with living things. IoT is making a critical imprint in numerous fields. These days, the versatile idea of IoT has changed and a conventional client can use it. IoT has created a few methodologies that make man's life simpler and comfortable. Aside from man's solaces, these strategies ought to be executed on necessities like food, which is accomplished from the farming fields. World Bank assessed that an overabundance is to be delivered before 2050 if the populace pattern is at the present rate. However, the current environment changes wouldn't sustenance such enormous yield creation. So sensors related to field, drones, progressed work vehicles and aquaculture cultivating may assist future ranchers with yielding more harvest at low costs. Like this, the need for exquisite cultivating is developing dramatically. The blend of Savvy Irrigation and the control to ML algorithms can find various disputes of agriculture. In ML adventure, the principle things need to be specific about the dataset and the algorithms. This book discusses the assistance of IoT and machine learning in farming, which can expand the productivity of yield production. Different climate parameters are taken into thought from which the best reasonable yield is grown predicted by a supervised learning algorithm, decision tree, along with classifier also discuss various ML applications in which rural ranches can be broadly utilized in regions like sickness recognition, crop discovery, irrigation system, soil conditions, also the Product quality and market analysis. By reading this book, readers can find the many noble aspects such as:

- *Machine Learning and IoT in Agriculture*
- *Precision farming and its application*
- *Precision farming in modern agriculture*
- *ML-based smart farming using LSTM*
- *Smart Weather Monitoring System using Sense Hat for improving the Quality of Crops*
- *IoT enabled smart farming: Challenges and Opportunities*
- *Fermat Point-Based Wireless Sensor Networks*
- *Application of IoT Enabled with 5G Network in Agricultural Sector*
- *An Economical Helping Hand for Farmers- Agricultural Drone*
- *Automatic Hibiscus Leaf Disease Detection and Classification Using Unsupervised Learning Techniques*
- *On Securing Smart Agriculture Systems: A Data Aggregation Security Perspective*
- *Urea Spreaders for improving the Crop Productivity in agriculture: Recent Developments*
- *Agricultural Informatics and Practices*

In summary, in this twenty-first century, there is a genuine need for agriculture upgradation. This book would provide a technical overview that leads to open a new dimension that may be useful to cover the solutions of the current growth of the agriculture process.

Dehradun, India — Amitava Choudhury
Asansol, India — Arindam Biswas
Dehradun, India — T. P. Singh
Asansol, India — Santanu Kumar Ghosh

# Contents

# Editors and Contributors

## About the Editors

**Amitava Choudhury** is an Assistant Professor in the School of Computer Science, University of Petroleum & Energy Studies, Dehradun, India. He received his M.Tech. degree from Jadavpur University and completed his Ph.D. from the Indian Institute of Engineering Science and Technology, Shibpur. He has over eight years of teaching and two years of research experience. His areas of research interest are computational geometry in micromechanical modeling, pattern recognition, character recognition, and machine learning.

**Arindam Biswas** is an Associate Professor in School of Mines and Metallurgy at Kazi Nazrul University, Asansol, WB, India. He received his M.Tech. degree in Radio Physics and Electronics from the University of Calcutta in 2010 and a Ph.D. from NIT Durgapur in 2013. Dr. Biswas has 12 years of teaching, research, and administrative experience. He has 55 journal papers, 35 conference proceedings, 07 authored books, 07 edited books, and 06 book chapters to his credit. Dr. Biswas has supervised 05 Ph.D. students in different topics of applied optics and high-frequency semiconductor devices. His research interest areas are carrier transport in the low dimensional system and electronic device, non-linear optical communication, and THz semiconductor source. Dr. Biswas served as a reviewer for reputed journals, a member of the Institute of Engineers (India), and a regular fellow of the Optical Society of India (India).

**T. P. Singh** is a Professor and Head of the Department of Computer Science, University of Petroleum & Energy Studies, Dehradun. Dr. Singh holds a Doctorate in Computer Science from Jamia Millia Islamia University, New Delhi. Dr. Singh has 25 years of academics, administrative, and industrial experience. His research interests include machine intelligence, pattern recognition, and the development of hybrid intelligent systems. To his credit, he has over 50 publications in national and international journals. He has guided 15 master's theses and is currently supervising 06 doctoral candidates.

**Santanu Kumar Ghosh** received his B.Sc. and M.Sc. degrees from the University of Calcutta, in 1996 and 1998, respectively. He obtained his Ph.D. degree from Jadavpur University, in 2006. Prof. Ghosh has 19 years of teaching experience. His areas of research are production planning, inventory management, and supply chain management. He has supervised 2 Ph.D. students and is currently guiding 6 Ph.D. students. He has published several research papers in international journals.

## Contributors

**Arwa Al-Turki** Riyadh, Saudi Arabia

**Ghada Alateeq** Riyadh, Saudi Arabia

**Tala Almashat** Riyadh, Saudi Arabia

**Nora Alqahtani** Riyadh, Saudi Arabia

**M. Appadurai** Department of Mechanical Engineering, Dr. Sivanthi Aditanar College of Engineering, Thiruchendur, Tamil Nadu, India

**Anees Ara** Riyadh, Saudi Arabia

**K. Athiappan** Department of Civil Engineering, Jyothi Engineering College, Thrissur, Kerala, India

**Kirti Panwar Bhati** School of Electronics, Devi Ahilya University, Indore, India

**Arindam Biswas** School of Mines and Metallurgy, Kazi Nazrul University, Asansol, India

**Ratan Das** National Research Centre for Grapes, Pune, India

**Kaushik Ghosh** School of Computer Science, UPES, Dehradun, Uttrakhand, India

**Harshita Jain** School of Electronics, Devi Ahilya University, Indore, India

**Ravish Jain** Mechanical Engineering Department, Birla Institute of Technology Mesra, Ranchi, Jharkhand, India

**Supriya Jaiswal** Department of Electrical Engineering, National Institute of Technology, Hamirpur, HP, India

**Virendar Kadyan** Department of Informatics, School of Computer Science, University of Petroleum and Energy Studies, Dehradun, India

**Nupoor Katre** School of Electronics, Devi Ahilya University, Indore, India

**Deepika Koundal** Department of Systemics, School of Computer Science, University of Petroleum and Energy Studies, Dehradun, India

**Mainak Mandal** Mechanical Engineering Department, Birla Institute of Technology Mesra, Ranchi, Jharkhand, India

**Prashant Meshram** School of Electronics, Devi Ahilya University, Indore, India

**Kaushal Mukherjee** Department of Electronics and Communication Engineering, National Institute of Technology, Jote, Arunachal Pradesh, India

**Subhadeep Mukhopadhyay** Department of Electronics and Communication Engineering, National Institute of Technology, Jote, Arunachal Pradesh, India

**Siqabukile Ndlovu** Computer Science Department, National University of Science & Technology, Bulawayo, Zimbabwe

**Sindiso M. Nleya** Computer Science Department, National University of Science & Technology, Bulawayo, Zimbabwe

**Aman Pandey** Mechanical Engineering Department, Birla Institute of Technology Mesra, Ranchi, Jharkhand, India

**Devendra Pandey** Central Institute for Subtropical Horticulture, Uttar Pradesh, Lucknow, India

**Himanshu Pandey** YSP UHF Nauni, Solan, Himachal Pradesh, India

**Richa Pandey** Mechanical Engineering Department, Birla Institute of Technology Mesra, Ranchi, Jharkhand, India

**P. K. Paul** Department of CIS, & Information Scientist (Offg.), Raiganj University, Raiganj, India

**E. Fantin Irudaya Raj** Department of Electrical and Electronics Engineering, Dr. Sivanthi Aditanar College of Engineering, Thiruchendur, Tamil Nadu, India

**Gopal Rawat** Department of Electronics and Communication Engineering, National Institute of Technology, Hamirpur, HP, India

**Reek Roy** Department of Computer Science, Belda College, Vidyasagar University, Belda, West Bengal, India

**Sahadev Roy** Department of Electronics and Communication Engineering, National Institute of Technology, Jote, Arunachal Pradesh, India

**Himadri Nath Saha** Department of Computer Science, Surendranath Evening College, Calcutta University, Kolkata, West Bengal, India

**Sugandha Sharma** School of Computer Science, UPES, Dehradun, Uttrakhand, India

**Sumiksha Shetty** Department of ECE, Sahyadri College of Engineering & Management, Mangalore, India

**Devendra Singh** Motilal Nehru National Institute of Technology, Uttar Pradesh, Allahabad, India

**A. B. Smitha** Department of ECE, Sahyadri College of Engineering & Management, Mangalore, India

# Chapter 1
# Smart Agriculture Using IoT and Machine Learning

**Sumiksha Shetty and A. B. Smitha**

**Abstract** Agriculture is the essential and most vital profession of our nation because it balances food necessity and furthermore the fundamental crude materials for a few enterprises, and there is a consistent expansion sought after with populace development. Various everyday issues are getting an incredible effect by the progressions in innovation. From a few years, individuals are running after the robotization with some degree of knowledge to supplant or limit human from the cycles of process. Innovation in agriculture lessens reliance on individual human work and land. The innovation permits operational concocting and speeds up decision-making on the farms. Ongoing headways in innovation have an extraordinary effect on agriculture and it has been set up that IoT is utilized in cultivating to improve nature of farming. Advancement of Machine Learning (ML) and Internet of Things (IoT) has assembled consideration of specialists to apply these strategies in fields like farming. It causes ranchers to expand the profitability of their property so the overall demand for food can be satisfied. IoT is a high-level innovation for observing and controlling gadget in any place on the planet. It can interface gadgets with the living things. IoT is making a critical imprint in numerous fields. These days, the versatile idea of IoT has changed; IoT can be used by a conventional client. A few methodologies that IoT has created make man's life simpler and comfortable. Aside from man's solaces, these strategies ought to be executed on essential necessities like food, which is accomplished from the farming fields. World Bank assessed that an overabundance to be delivered before the 2050 if the populace pattern is at present rate. However, the current environment changes wouldn't sustenance such enormous yield creation. So sensors which are related to field, drones, progressed work vehicles and aquaculture cultivating may assist future ranchers with yielding more harvest, at low costs. Thusly, the need for exquisite cultivating is developing dramatically. The blend of Savvy Irrigation and the control to ML algorithms can find various disputes of agriculture. In ML adventure, the principle things need to be specific about the dataset and the algorithms. In this chapter, we discus about the assistance of IoT and machine learning in farming, which can expand the productivity of yield production. Different climate parameters are taken into thought from which the best reasonable yield is grown is predicted by

S. Shetty (✉) · A. B. Smitha
Department of ECE, Sahyadri College of Engineering & Management, Mangalore, India



A. Choudhury et al. (eds.), *Smart Agriculture Automation using Advanced Technologies*, Transactions on Computer Systems and Networks,
https://doi.org/10.1007/978-981-16-6124-2_1

supervised learning algorithm, decision tree, along with classifier also discuss about various ML applications in which rural ranches can be broadly utilized in regions like sickness recognition, crop discovery, irrigation system, soil conditions and also the product quality and market analysis.

**Keywords** Machine learning · IoT · Decision tree · Supervised learning algorithm

## 1.1 Introduction

Farm management and agriculture suggests a significant part in Gross Domestic Product not just of developing republics yet additionally for few industrialized republics. Consequently, extemporizing and also progression of the existing agricultural improvements are of excessive significance. It is not solitary backing in flourishing supportable enhancement of mortality, broadly speckled foliage be that as it may, will support in handling the universal alternative, likely, ecological variation and scourges like flow. With improved modernization comes improved yield; subsequently, it aids forestall surroundings like malnourishment and ailing health. The modernization ought to be manageable at a moderate rate and its outcome possibly will spread to millions of personalities everywhere in the world (Maduranga and Abeysekera 2020). In this cutting-edge world, the greater part of the rancher needs appropriate information with regard to cultivating and farming making it more sporadic. Most of the cultivating- and farming-related exercises depend on expectation and determining. At the point when it comes up short, the ranchers need to bear tremendous misfortunes and some wind up ending it all. Since we know about the advantage of nature of soil, air, irrigational and in the development of yields such boundaries like temperature, moistness can't be ignored. According to a study done by United Nations—Food and Horticulture Organizations, the overall food creation ought to be expanded for about 70% in the year 2050 for developing populace. Horticulture is the fundamental wellspring of food and it assumes significant part in the development of nation's budget. In addition to this it gives enormous adequate effort freedoms to individuals. The ranchers are still utilizing conventional strategies for farming, which brings about the yielding of harvests and natural products down. So the harvest yield can be upgraded by utilizing programmed apparatuses (Elijah et al. 2018). It is essential to accomplish existing skill and modernization in the farming for expanding the harvest. The mix of conventional strategies with most recent advances leads to the modernization in agriculture which utilizes Internet of Things and machine learning. As of late, IoT is been utilized to address the difficulty of various mechanical and specialized purposes. Presently, it is an ideal opportunity to satisfy the need of future cultivating which must be refined by brilliant Agro-IoT device. IoT helps in persistent observing of the field to give valuable data to the ranchers which will enhance another period in future cultivating. IoT device can be executed for checking environmental variation, water management, terrestrial observing, expanding profitability, observing harvests, controlling pesticides and soil management, distinguishing plant sicknesses, expanding the pace of yield deal and so on IOT is an arising subject of specialized, social and reasonable turn of events (Maduranga and Abeysekera 2020). Objects identical to consumer things, automobiles, huge apparatuses, automatic and utility sector are associated with web accessibility giving essential data that assurance to alter the way where we work simplifying our life. The use of IoT in agribusiness is tied in with enabling ranchers with the choice devices and computerization advancements that consistently incorporate items, information and administrations for better efficiency, quality and benefit. Ongoing reviews on the IoT in agribusiness have zeroed in on the difficulties and limitations for huge scope pilots in whole production network in agri-food area. The blend of Savvy Irrigation and the control to ML algorithms can find various disputes of agriculture. In ML adventure, the principal things need to be specific about the dataset and the algorithms. Advancement of Machine Learning (ML), and Internet of Things (IoT) has assembled consideration of specialists to apply these strategies in fields like farming (Elijah et al. 2018).

## *1.1.1 IoT Ecosystem*

It comprises four significant segments:

- IoT gadgets.
- Communication technology.
- Internet.
- Information stockpiling and preparing (Fig. 1.1).

### 1.1.1.1 IoT Gadgets

The IoT gadgets comprise software frameworks which cooperates with devices which entails wireless availability. These gadgets would now and then be alluded as IoT sensors. This is utilized to screen and to quantify diverse ranch factors and also factors which influence creation. The devices are classified to different area, optical, power driven, and also the electrochemical devices (Ersin et al. 2016). From this data is been assemble, for example, air, soil temperature for different profundities, precipitation, leaf humidity, chlorophyll, speed, wind heading, sun-powered radiation and air pressure.

### 1.1.1.2 Communication Technology

The communication modernization adopts a critical part in effective arrangement of the Internet of Things outlines. The contemporary communication modernization is grouped reliant on standard band and appeal. This communication is the collection of

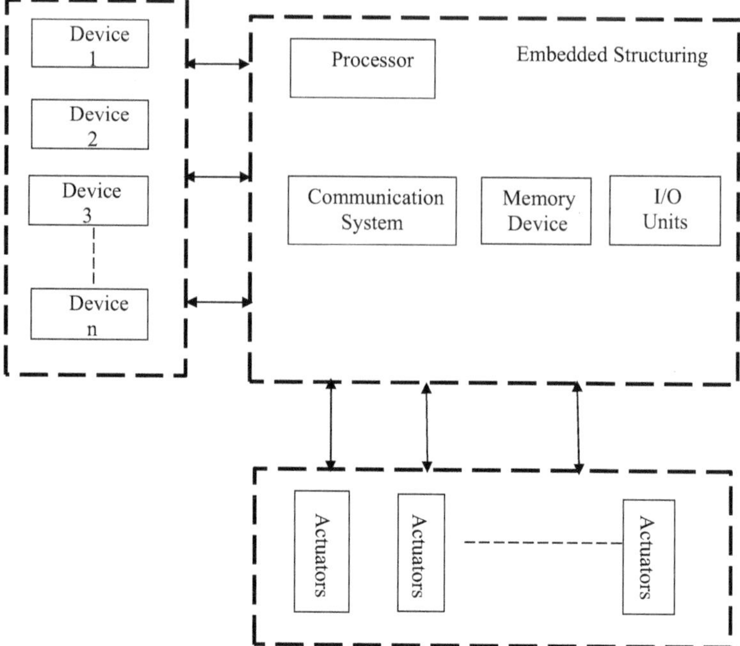


**Fig. 1.1** Structure of IoT ecosystem

short-range correspondence standard, furthermore long-range correspondence standard (Maduranga and Abeysekera 2020). The range of correspondence is amassed to be sanctioned along with the unlicensed spectrum. These contraptions application states are originated in device or in the backhaul organization, deployment conditions.

#### 1.1.1.3 Internet

Headway in the arena of remote communiqué frameworks, cell phones and omnipresent administration is clearing the route for gigantic availability of the Internet. As indicated by Machina research, quantity which is associated with agronomic gadgets would predict that it is developed from 13 million towards the finish of 2014 and also at the end of 2024 it would be 225 million. Internet shapes the centre web layer, where ways given to convey and trade information and organization data among different subarrays. The association of Internet of Things contraptions with the net empowers information that can be accessible from everyplace and whensoever. Exchanging information by means of Internet requires sufficient security, backing of constant information and simplicity of availability. Internet has a cleared route for distributed computing, where huge information is accumulated for capacity and processing. Distributed computing includes the administration of UI, organizing along with the directing of web hubs, figuring and preparing information.

Towards the accomplishment of the availability of diverse frameworks and gadgets over Cyberspace, connection between hardware layer and application layer is identified as IoT middleware. This is accountable for interaction among the devices and information regarding management system and hence IoT middleware and network conventions are being created. Instance for IoT middleware is SOA which is service-oriented architecture, actor and cloud-based IoT middleware is pragmatic to help IoT. The SOA comprises multi-facet design for IoT (Bacco et al. 2018). A portion of the recommended IoT architecture comprise the accompanying layers: detecting, getting to, organizing, middleware and the application layers.

#### 1.1.1.4 Information Stockpiling and Preparing

Information-driven agribusiness includes the assortment of huge, dynamic, multi-faceted and spatial data, stowage and handling. The intricacy of the information can go from organized to nonstructured information, which could be text, pictures, sound and audio-visual. The information obtained is from recorded information, device information, commercial and also souk-associated information (Maduranga and Abeysekera 2020). Utilizing the cloud IoT stages takes into account large information gathered from sensors to be put away in the cloud. This incorporates facilitating of utilization that is basic in offering types of assistance and to oversee end-to-end communication of IoT architectonic. As of late, authority or mist registering is pushed, whereas in the IoT devices and passages complete calculation and examination to lessen idleness for basic applications diminish cost and advance QoS. There are a few agribusinesses the executives' data frameworks that have been created to deal with the different types of information.

In the organization of IoT gadget, some of the key technical boundaries should be thought of. For remote availability, the accompanying boundaries ought to be thought of the scope of correspondence distance, information rate, battery life, portability, dormancy, security and flexibility, and the price of the gateway data processor.

Among the correspondence innovation, LPWA would be drawing a lot of interest particularly arising from NB-IoT. This guarantees stimulating the highlights along with low gadget current utilization, ultralow gadget cost, laid-back to contrivance, backing from an enormous quantity of low-throughput gadgets, significant detachment inclusion which could uphold transfer as well as downloading of information (Vij et al. 2020). Few practice instances of the IoT which have been embraced by agriculture.

### *1.1.2 IoT Applied in the Field of Agriculture*

There are a few instances of use of IoT in the agronomy. Instances of these cases are yield as well as livestock, water system and water quality observing, climate checking, soil observing, infection and bug control, computerization and accuracy. The use of IoT discusses about dependency on the accompanying capacities: checking, farming apparatus, exactness agribusiness and nursery creation.

#### 1.1.2.1 Monitoring

In farming, a few variables can be checked, and these elements rely upon the area of agriculture underneath contemplations. The vital variables which must be observed are featured and also examined.

Yield Farming

In yield cultivating, a few ecological elements influence ranch produce. Securing such information support to comprehend the examples, also homestead interaction. This information incorporates the measure in precipitation, greenery vapour, heat, dampness, mud dampness, saltiness, environment, dry circle, sun-oriented radiation, bug development, human exercises and so forth.

The securing of such point-by-point record empowers ideal choice making in order to improve the nature of the ranch produce, limit hazard and amplify benefits. For example, the sun-powered radiation information gives data approximately the openness floras to daylight as of, since the rancher could recognize if floras remain appropriately uncovered or else over uncovered (Maduranga and Abeysekera 2020). The mud sogginess eases permit data upon the sogginess of mud, and monitoring mud situations also lessens the danger of shrub illnesses which can be controlled. Besides, opportune and exact climate gauging information, for example, climatical variations also precipitation, can progress the efficiency degree. Moreover, this information can support ranchers in arranging organize, also decrease ranchers' expense of work. The ranchers could likewise arrange restorative and pre-emptive procedures in advance of period at the mercy of on the statistics given. The nuisance blossoming statistics can be congregated and distantly took care of conscious for the ranchers to bug regulator, otherwise pre-owned to give exhortation in the direction of the rancher's dependency on evidence tracing of vermin assaults.

Aquaponics

Aquaponics is blend of aquaculture along with hydroponics, where the fish squanders are taken care of into plant homesteads to give the fundamental supplements needed via the floras. This defines homesteads, the thing is critical in the direction of continually screening the condition of liquid, heat and water degrees, soundness of the aquatics, pH degree, moistness and also the sunbeams. This exact information could also progress the yield of aquatics and florae since it permits essential nourishment for the growth between the florae and aquatics. This information could likewise utilize aimed at robotization tenacities where anthropological intercession would be less.

Forestry

Ranger service assumes a significant part in carbon cycle, which is furthermore harbour more than 66% of the world species known. The components to be checked in forest incorporates are temperatures and dampness of air and soil, and the various degrees of gases, like carbon monoxide, $CO_2$, toluene, $O_2$, $H_2$, methane, $C_4H_{10}$, smelling salts, ethanol, $H_2S$ and $NO_2$ (Maduranga and Abeysekera 2020). These boundaries provide early admonition and ready frameworks in contradiction of plain firing under the backwoods, also furthermore support to screen in contrast to infections.

Livestock Cultivating

The elements which need to be observed in animals rely on the kinds in creatures viable. For instance, the potential in the milk from wild oxen, also bullies provide data approximately the well-being condition of the creatures. Different variables are heat, mugginess, harvest, bug assault and also the water quality. The arrangement and execution arrangement likewise permit ranchers to trail and enquire the area of their domesticated animals by labelling singular creature with RFID gadget, subsequently forestalling creature robbery.

Different territories, for example, stockpiling observing which incorporates water, and fuel, along with creatures' feeds can likewise be observed, and the information can assist the ranchers with preparing and saving cost.

While a few arrangements given as the space in checking, selection in minor and with the average gauge ranches are actually restricted particularly around agricultural nations because of need of mindfulness and arrangement cost. The possibility to create practical horticultural base IoT arrangements is as yet a very open territory.

#### 1.1.2.2 Trailing and Locating

IoT could likewise be spread in resource following to progress organization's production network and coordination. IoT can give information to empower agrarian organizations to settle on better decisions, arranging, wisely interfacing with its commercial partners and set aside time and cash. Data like area, resource distinguishing proof can be followed utilizing RFID and also cloud-based worldwide aligning framework. Trailing and following in farming item bind permit the shopper to know about the total history of an item, in this way improving the purchaser's trust regarding the item well-being and well-being-related issues. Though tracking is the capacity to catch, wrinkle and stock knowledge acknowledged with the range link from the material inputs needed for production to the products produced and distributed., tracking licenses the things to be predictable from the products produced to the material inputs needed for production (Ersin et al. 2016), which also permits a limited evidence to be grouped in addition to the catalogue link with the end goal that the consumer and diverse partners have been ensured upon the inception, area and also item's life history.

There are a few aspects which could be followed and also incorporate the developing climate, production circumstances, pest factors, the executives' factors, stockpiling conditions and also the transportation, stretch to souk. This element can equally extant abrupt or probable well-being hazard to buyers. These critical variables also influence the emerging environment which can be soil, airborne and also the liquid. This situation is affected because of the use of weed killer, composts and insect repellent. Also, the sort of forages, also immunizations controlled via animals could be followed, in the meantime may straightforwardly affect well-being security problems. Agrarian items can influence to bother along the whole cycle, which could influence the amount and nature of the item, following the items can assist the ranchers with improving the store chain along with the production.

A following and locating framework has to be essentially consisting of statistics input, hoarding, transfer, rotation and harvest. This input information integrates the information in the whole lifespan of the item, basis, location currently located, terminus and also the partners engaged with the whole production network. The frameworks have to similarly integrate reminiscence in order to stockpile the statistics throughout some stretch of time aimed at advanced exertion tenacities. The statistics change suggests to the mode to binding collectively and standardizing the whole statistics. This framework ought to likewise have the option to deal with the information gathered lastly yielding it to everybody required along the store network. The utilization of RFID to track since the creation point, preparing, shipping, stockpiling, circulation, deals and later deal administrations are featured. It bounces the measurements to pleat, stock and also investigate statistics on a significant detachment rapidly.

#### 1.1.2.3 Agricultural Machinery

Agricultural apparatus based on IoT can aid to improve the crop profitability and decrease grain misfortunes. By appropriate planning and with the help of GPS and worldwide route satellite frameworks (GNSSs), the hardware can be worked in autopilot method.

This machine may incorporate vehicles (UAVs), i.e. automated airborne vehicles and the automatons could also distantly be structured depending upon the accessible data gathered by means of the IoT framework for accurate and operative use of assets to required homestead zones. The hardware can likewise gather information and this information could support ranchers to plan their arena for arranging line-ups, for example, impregnating, water system and sustenance (Maduranga and

Abeysekera 2020). For model, CLAAS, an agrarian hardware maker has executed IoT on their hardware, empowering their machinery to be worked by making use of autopilot system. Another arrangement is Precision Hawk's UAV devices, which can give ranchers data, for example, wind speed, pneumatic stress, among other boundaries. The arrangement can likewise be utilized for symbolism and also planning of agricultural conspiracies.

#### 1.1.2.4 Precision Agriculture

Exactness horticulture can essentially be characterized as the assortment of constant information from ranch variables and utilization of prescient examination for brilliant decisions to augment yields, limit natural effect and decrease of cost accuracy agribusiness depends on different innovation which incorporate sensor hubs like GPS and large machine learning (ML) to accomplish improved harvest yield. The savvy choice accomplished from the ML additionally brings about less misuse of assets, like water in water system frameworks, manure, pesticides and so forth. Accuracy agribusiness grants new-fangled difficulties for specialists fashionable to the extent of mechanical technology, climatological information detecting, also image processing and so on. With GPS along with the GNSS, ranchers can find exact area and guide locales with a few information factors, which would be then utilized by factor rate innovation to ideally circulate ranch assets, for example, cultivating, crop-dusting and different administrations. In spite of the fact that accuracy agribusiness innovation can also improve the yield, the fundamental to give solution that would not be difficult to practice by means of the ranchers also furthermore gives exercise towards empowering little and intermediate gauge ranchers profit by the frameworks. Precision agribusiness can fundamentally be described as an arrangement of continuous statistics on or after ranch variable quantity and usage of perceptive assessment towards shrewd verdicts to enhance vintages, limit natural impact and abatement cost (Ryu et al. 2015).

#### 1.1.2.5 Greenhouse Production

Nursery otherwise called the glasshouse innovation which is a procedure, where floras are full-fledged under-controlled climate. The advantage is of developing any floras in any spot whenever by giving appropriate ecological conditions. A few contemplates have been done on the use of the WSNs in nursery to screen ecological conditions. Ongoing works have revealed how the application of IoT on greenhouse can be applied so that it decreases human asset, save energy, increment proficiency in nursery site checking and direct association of nursery ranchers to clients (Maduranga and Abeysekera 2020).

### *1.1.3 Machine Learning Applications in IoT-Based Agriculture*

ML is considered as a new route for machines to mimic people learning exercises, acquire new information, persistently improve execution and accomplish one-of-a-kind developments.

In recent years, ML is one of the extremely fruitful in calculations, hypotheses and applications, which has been joined with other farming procedures to limit the crop costs and also boost the yield. ML applications on agrarian ranches can be generally utilized in regions like infection recognition, crop location, water system arranging, soil condition and weed discovery along with crop quality and climate anticipating. ML for dissecting the newness of harvest is found after reap, shelf life, souk examination, also product quality, and so forth.

#### 1.1.3.1 Plant Management

Amalgam of ML along with the IoT gives suitable and untroublesome climate to developing yields through nursery innovation. Nonetheless, the spatio-transient inconstancy of yield development of ecological boundaries and their common impacts in ensured agribusiness make it hard for conventional horticulture and natural guidelines to adjust to the development of various sorts of floras at various phases of development. Consequently, higher exactness is required from a checking and control point of view. Numerous works exist on planning and testing kinds of observing and control frameworks for changing temperature and dampness, splendour, $CO_2$ fixation and other natural boundaries for the IoT, specialized and financial outcomes (Elijah et al. 2018). It is recommended that regulating the climate conditions for a particular sort of shrub can be meticulous through IoT, feelers and actuators. The guidelines of controlling can be done by an Artificial Neural Network (ANN) arrangement for IoT cloud.

#### 1.1.3.2 Crop and Yield Management

ML-based harvest planning is applied in ranches dependent on gathered information over Internet of things over yield observing associated through GPS. The gathered date which uncovers the yield subtleties will be planned dependent on the kinds of homestead land. Aside from that, machine learning framework along with IoT has been used to foresee and also improve the harvests in horticulture. Ranchers depend principally on rural specialists to decide. Ranchers and others utilize these frameworks with no information on PC use. ML framework can be utilized for yield production (Maduranga and Abeysekera 2020). This is an information building framework that produces data utilizing existing information.

This empowers ranchers to make financially stable yield the management choices. Different such frameworks have been created considering the accomplishment of master frameworks. The Internet IoT assumes a significant part in farming. Related works in ML frameworks can be fabricated on the IoT and can make proposals on the utilization of information gathered continuously.

#### 1.1.3.3 Soil Management

An ML-based methodology is applied for the management of soil. Soil information can be gathered from remote sensor hubs sent on site.

At that point, gathered information can be taken care for ML calculations to anticipate and break down soil properties or arrange the kinds of soil utilizing administered ML calculations. In addition, most generally utilized ML calculations, K-closest neighbour, support vector relapse (SVR), Naive Bayes and so on can be used to anticipate soil dryness dependent on rainfall and also evaporative hydrology information.

#### 1.1.3.4 Diseases Management

ML along with IoT has been utilized to distinguish and oversee infections in rural fields. ML strategies further invigorate proper pesticides to shield crops from the diseases what's more, decrease work. Such framework helps makers by acquiring measurements and arranging manures, pesticides and water system likewise. By precisely recognizing the infection and giving precise pesticide application and water system plans, grape perceivability and volume have been expanded and outrageous pesticide utilization has been decreased. Moreover, design with profound learning strategies for recognizing and arranging discourse steps of different plants. The sound strides at these plants are in light of ongoing caught visual data and travel through various spaces of the ranch by means of IoT-based sensor hubs sent in yield field.

#### 1.1.3.5 Weed Management

Management of weed is basics for any cultivating. Weed planning through ML has been researched. To streamline this, propose automated flying machine in order to capture pictures and guide weed in an arena. Here flying machine has been controlled via IoT organization. Progressed IoT innovations, for example, NB-IoT is utilized to deal with and control huge amount of information.

#### 1.1.3.6 Water Management

A few frameworks have been executed on monitoring the supply of water for a horticulture field just as dissecting the water value utilizing ML. It tends to be created astute frameworks that distinguish ground boundaries, for example, soil dampness, soil temperature and natural conditions utilizing IoT sensors. At that point, utilize a similar information to anticipate outdoor relative dampness. Moreover, we can utilize half breed ML and IoT frameworks for the control of water temperature and conform to encompassing temperature in insightful manner.

### *1.1.4 Several Methods Have Been Incorporated in Machine Learning Which Has Been Listed Below*

#### 1.1.4.1 Prediction

IoT gives enormous evidence that could be concentrated subsequently around little time to judge the existing biological situations. This across various sorts of organizations sensors could stay expected employing ML and keen intention can be developed in anticipating the ecological changes which also provide statistics obsessed measures.

In spite of this circumstance, the IoT information is used in monitoring dissimilar portions of a ranch, such as the aquatic structure contexts, and the evidence can similarly employ to forecast furthermore thoughtfulness landowners in contradiction of sickness or outrageous environment situations, like flood or dry season. For example, in forestry, the devices can be employed to awning fire episode or foresee to locale in a forest that stretches great hazard of fire flare-up. These statistics could support the firemen with captivating precautionary procedures towards the specific extent (Elijah et al. 2018). Additional space of prospect integrates initial admonition in contradiction of cataclysmic proceedings to progress catastrophe response.

#### 1.1.4.2 Storage Management

Enormous number of farming items are typically lost because of helpless stockpiling of the board framework. Whereas heat, dampness and additional natural aspects significantly influence the adulteration in nutrition items, creepy crawlies, microorganism, gnawer and so on can influence the quality and amount of the food items. The utilization of IoT along with ML for stowage frameworks can assist with improving horticultural item stockpiling. Sensors are conveyed to screen the stowage spaces and also ecological situations. This information is directed for the haze, also examined. The auto-robotized choice framework which depends on means of the examined information could be sent in the direction to change the ecological settings. Additionally, an admonition alarm could be started to ranchers once outrageous circumstances are grasped or else if vermins are revealed in the storeroom. In Republic of India, for that around 40% to 45% of the new item vanished, and subsequently, later, the reaping because of a few components which incorporates decay of a cold stockpiling the board framework is planned based on IoT innovation, where the storeroom works under the controlled temperature. In spite of the fact that IoT can progress the agronomic storage space, security ought to be installed into such framework to keep away from item robbery.

#### 1.1.4.3 Decision

The information has been collected from the sensor which is required to make a decision. The huge information acquired from device bids learning opportunities so that dynamic in continually changing natural conditions been improved, and this can be either short medium or in long terms. Computerized choices can be produced using the Internet of Things framework when firm circumstances are grasped, accordingly needing of fewer or may be on no account of humanoid mediations. This mechanized choice can be regulating the temperatures in order to supply water in a controlled manner from a water system framework. For example, in nurseries, the utilization of ML can help to decide the ideal conditions for which to cultivate a specific harvest by noticing the information obtained from sensors identifying with supplements, yield, development, happening, shading, taste and re-transplantation, furthermore air quality. Strategy settling on choice by regime and also the investors can likewise be upgraded by the measure of information acquired from ML, thusly it is significant that the information is exact, compact, total and on schedule.

A few rural dynamic frameworks have been created to empower ranchers to settle on educated choice with regard to their homesteads and domesticated animals. The ML gives choice on technical direction to ranchers, vermin and sicknesses control and proposal from distant master direction frameworks.

#### 1.1.4.4 Farm Management

Cohesive ranch management framework permits a whole homestead to be checked. Information is gathered through an organization of sensors counting along with the body devices which, in turn, are creatures through a solitary reason of pouring efficiency. Three significant components which incorporate hazard managing, rate and also efficiency harvest should oversee through ongoing data and appropriately advanced to augment efficiency.

ML assumes a significant part to introduce the ranchers, partners with huge information that can be painstakingly examined to stay away from pointless danger or execute precautionary measures to improve efficiency.

DA additionally permits different homesteads to be associated and overseen on a solitary stage, where data on logical advances, creation, advertising, ranch management, suggestions and other correlated themes are spread to expand profitability, yield and income.

#### 1.1.4.5 Precise Application

ML utilizing estimated information from devices can empower exact utilization of synthetic substances and composts to explicit spaces of the ranch, and this can improve the efficiency while diminishing the cultivating cost. In spite of the fact that accuracy cultivating frameworks have been sent in ranches in cutting-edge nations, agricultural nations are starting to embrace the innovation particularly in research ranches. In any case, the organization cost, innovation and mindfulness actually restrict the organizations of the IoT-based accuracy cultivating frameworks in agricultural nations. Moreover, the greater part of the homesteads in non-industrial nations are little scaled ranches, and most ranchers don't apply such expertise. Creating proper exactness cultivating arrangements for little homesteads actually stays an open region for specialist and architects. Another benefit of ML in accuracy cultivating is its application in directing apparatus utilizing GPS and area information to exact areas in the homestead in this way improving the cultivating productivity when contrasted and hominoid-driven machineries. This not only saves time, but also fuel and operational expense.

#### 1.1.4.6 Insurance

Ranchers are generally presented to outrageous climate conditions which might prompt helpless reap. Be that as it may, with the execution of IoT innovation ranchers can be safeguarded with their harvests and animals. An organization of devices can be sent; furthermore, observing can be accomplished by far-off automated stations. The information can be shipped off the cloud and examined.

The protection strategy can be installed with a notice framework, where outrageous climate conditions are anticipated and the protected ranchers are alarmed by instant messages. This can empower the ranchers adopt preparatory strategy to secure their homesteads. An additional benefit of ML in protection is the way the insurance organizations approach the information from the far-off ranches, which can start a computerized disbursement believed the Internet of Things versatile instalments frameworks once outrageous circumstances are noticed. This could wipe out the requirement aimed at protracted case measure for which the assurance agency desires to learn the degree of harm via staying the homesteads.

IoT-ML-based agribusiness is the following evolutional thing in shrewd agrarian and keen cultivating. Applying ML calculations to information created from different contributions from ranches with the assistance of the agrarian IoT will make the framework more astute and give conclusive data and make prophecies. The existing

ML applications in farming, from cycle to results, each of which it with its peculiar qualities and shortcomings. Afterwards, on the grounds that the ML applications required constant information to prepare prescient calculations, ideas were made to execute new applications on the IoT. Homestead the executives' frameworks are advancing into the reality for which the AI has been applied to sensor information. The Artificial Insight (AI) framework is more extravagant.

## References

Maduranga MWP, Abeysekera R (2020) Machine learning applications in IoT based agriculture and smart farming: a review. Int J Eng Appl Sci Technol 24

Elijah O, Rahman TA, Orikumhi I, Leow CY, Nour Hindia MHD (2018) An overview of Internet of Things (IoT) and data analytics in agriculture: benefits and challenges. IEEE Internet Things J 5(5):3758–3773

Vij A, Vijendra S, Jain A, Bajaj S, Bassi A, Sharma A (2020) IoT and machine learning approaches for automation of farm irrigation system. Procedia Comput Sci 167:1250–1257

Ersin Ç, Gürbüz R, Yakut AK (2016) Application of an automatic plant Irrigation system based arduino microcontroller using solar energy. Solid State Phenom 251:237–241. Trans Tech Publications Ltd

Bacco M, Berton A, Ferro E, Gennaro C, Gotta A, Matteoli S, Paonessa F, Ruggeri M, Virone G, Zanella A (2018) Smart farming: opportunities, challenges and technology enablers. In: 2018 IoT vertical and topical summit on agriculture-Tuscany (IOT Tuscany). IEEE, pp 1–6

Ryu M, Yun J, Miao T, Ahn I-Y, Choi S-C, Kim J (2015) Design and implementation of a connected farm for smart farming system. In: 2015 IEEE sensors. IEEE, pp 1–4

Manrique JA, Rueda-Rueda JS, Portocarrero JMT (2016) Contrasting internet of things and wireless sensor network from a conceptual overview. In: 2016 IEEE international conference on Internet of Things (iThings) and IEEE green computing and communications (GreenCom) and IEEE cyber, physical and social computing (CPSCom) and IEEE smart data (SmartData). IEEE, pp 252–257

de Lima GHEL, e Silva LC, Neto PFR (2010) WSN as a Tool for Supporting Agriculture in the Precision Irrigation. In: 2010 sixth international conference on networking and services. IEEE, pp 137–142

Elijah O, Orikumhi I, Rahman TA, Babale SA, Orakwue SI (2017) Enabling smart agriculture in Nigeria: application of IoT and data analytics. In: 2017 IEEE 3rd international conference on electro-technology for national development (NIGERCON). IEEE, pp 762–766

Dlodlo N, Kalezhi J (2015) The internet of things in agriculture for sustainable rural development. In: 2015 international conference on emerging trends in networks and computer communications (ETNCC). IEEE, pp 13–18

Park E, Cho Y, Han J, Kwon SJ (2017) Comprehensive approaches to user acceptance of Internet of Things in a smart home environment. IEEE Internet Things J 4(6):2342–2350

Da Xu L, He W, Li S (2014) Internet of things in industries: a survey. IEEE Trans Ind Inf 10(4):2233–2243

Premsankar G, Di Francesco M, Taleb T (2018) Edge computing for the Internet of Things: a case study. IEEE Internet Things J 5(2):1275–1284

Oche OE, Nasir SM, Muhammed AH (2020) Internet of Things-based agriculture: a review of security threats and countermeasures

Odema M, Adly I, Wahba A, Ragai H (2017) Smart aquaponics system for industrial Internet of Things (IIoT). In: International conference on advanced intelligent systems and informatics. Springer, Cham, pp 844–854

Oliveira I, Cunha RLF, Silva B, Netto MAS (2018) A scalable machine learning system for pre-season agriculture yield forecast. arXiv:1806.09244

Siddique T, Barua D, Ferdous Z, Chakrabarty A (2017) Automated farming prediction. In: 2017 Intelligent systems conference (IntelliSys). IEEE, pp 757–763

Aliac CJG, Maravillas E (2018) IOT hydroponics management system. In: 2018 IEEE 10th international conference on humanoid, nanotechnology, information technology, communication and control, environment and management (HNICEM). IEEE, pp 1–5

Araby AA, Elhameed MMA, Magdy NM, Abdelaal N, Allah YTA, Saeed Darweesh M, Fahim MA, Mostafa H (2019) Smart iot monitoring system for agriculture with predictive analysis. In: 2019 8th international conference on modern circuits and systems technologies (MOCAST). IEEE, pp 1–4

Pandithurai O, Aishwarya S, Aparna B, Kavitha K (2017) Agro-tech: a digital model for monitoring soil and crops using internet of things (IOT). In: 2017 third international conference on science technology engineering & management (ICONSTEM). IEEE, pp 342–346

Ananthi N, Divya J, Divya M, Janani V (2017) IoT based smart soil monitoring system for agricultural production. In: 2017 IEEE Technological Innovations In ICT for agriculture and rural development (TIAR). IEEE, pp 209–214

Adedoja A, Owolawi PA, Mapayi T (2019) Deep learning based on nasnet for plant disease recognition using leave images. In: 2019 international conference on advances in big data, computing and data communication systems (icABCD). IEEE, pp 1–5

Sarvini T, Sneha T, Sukanya Gowthami GS, Sushmitha S, Kumaraswamy R (2019) Performance comparison of weed detection algorithms. In: 2019 international conference on communication and signal processing (ICCSP). IEEE, pp 0843–0847

Machado MR, Júnior TR, Silva MR, Martins JB (2019) Smart water management system using the microcontroller ZR16S08 as IoT solution. In: 2019 IEEE 10th Latin American symposium on circuits & systems (LASCAS). IEEE, pp 169–172

Mahdavinejad MS, Rezvan M, Barekatain M, Adibi P, Barnaghi P, Sheth AP (2018) Machine learning for Internet of Things data analysis: a survey. Digital Commun Netw 4(3):161–175

Singh C, Sairam KVSSSS, Harish MB (2018) Global fairness model estimation implementation in logical layer by using optical network survivability techniques. In: International conference on intelligent data communication technologies and Internet of Things. Springer, Cham, pp 655–659

Soans RV, Hegde A, Singh C, Kumar A (2017) Object tracking robot using adaptive color thresholding. In: 2017 2nd international conference on communication and electronics systems (ICCES). IEEE, pp 790–793

Jayaraman PP, Yavari A, Georgakopoulos D, Morshed A, Zaslavsky A (2016) Internet of things platform for smart farming: Experiences and lessons learnt. Sensors 16(11):1884

# Chapter 2
# Precision Farming and Its Application

**Himanshu Pandey, Devendra Singh, Ratan Das, and Devendra Pandey**

**Abstract** Precision agriculture is a current approach that utilizes digital technology to examine and optimize crop growth and development process. Precision agriculture technology came into existence in the mid-1980s; by applying this advanced technique, the application of fertilizers at varying levels, further blending is easily possible in fields, and its optimization can be done easily for different cultivated crops. In the current scenario, this methodology is accepted worldwide across many countries in different crops. There are different technologies that are involved in precision agriculture like sensors, GPS, software, and remote sensing that have immense potential to increase crop production. Precision farming has applications in the collection of data from the field, yield assessment, remote sensing, quality mapping, and variable Fertilizer dose application and heat maps development. This chapter deals with the instrument and technology utilized by precision farming to increase the output of agricultural crops.

**Keywords** Precision agriculture · GPS · Remote Sensing · Sensors

## 2.1 Introduction

The fast-growing population of humans across the world has also kept increasing the food demands for their survivability. It's a huge challenge for the government as well as for the farmers to fulfill the continuously increasing food demands with the restricted resources (Mumtaz et al. 2017). Some latest advanced technologies can be

H. Pandey
YSP UHF Nauni, Solan, Himachal Pradesh 173230, India

D. Singh (✉)
Motilal Nehru National Institute of Technology, Uttar Pradesh, Allahabad 211004, India

R. Das
National Research Centre for Grapes, Pune 412307, India

D. Pandey
Central Institute for Subtropical Horticulture, Uttar Pradesh, Lucknow 226101, India

 
A. Choudhury et al. (eds.), *Smart Agriculture Automation using Advanced Technologies*, Transactions on Computer Systems and Networks,
https://doi.org/10.1007/978-981-16-6124-2_2

integrated into the traditional agriculture field to enhance crop productivity which can be helpful in dealing with this problem.

Precision farming (PF) is the latest farming approach used by farmers. The precise amounts of inputs are employed to get better average crop yields in comparison with the traditional cultivation approach and practices. Therefore, PF is a novel system that is designed to enhance crop production by utilizing various aspects such as technology, management, and information, in order to increase productivity, improve crop quality, safeguard the environment and conserve energy (Shibusawa 2002). PF is consists of remote sensing methods using IoT (internet of things) sensors, which play a vital role by monitoring, acquiring, and providing the processed field crop-related data to the farmers (Abbasi et al. 2014). There are various parameters involved during the crop monitoring like temperature, water levels, soil parameters, and sunlight. These latest PF techniques help the farmers to specifically monitor the parameters that are required for growing a healthy crop and also provide the data that where and in what amount these parameters are required at a specific instance of the time period. All the collected data related to crops, such as weather condition, soil nutrients, presence of weeds and pests, are analyzed, and agronomic suggestions are being provided (Berntsen et al. 2006). Therefore, PF is an attractive concept that naturally fulfills the farmer's expectation of using the farming inputs more effectively, increasing profits, and environment-friendly high productivity. The recent advancements in PF today will provide nature-friendly agriculture technology for tomorrow. Particularly for the small farmers of developing countries, PF will bring a significant yield improvement with less input (Fountas et al. 2005). A diagrammatic representation of the PF cycle was shown in Fig. 2.1.

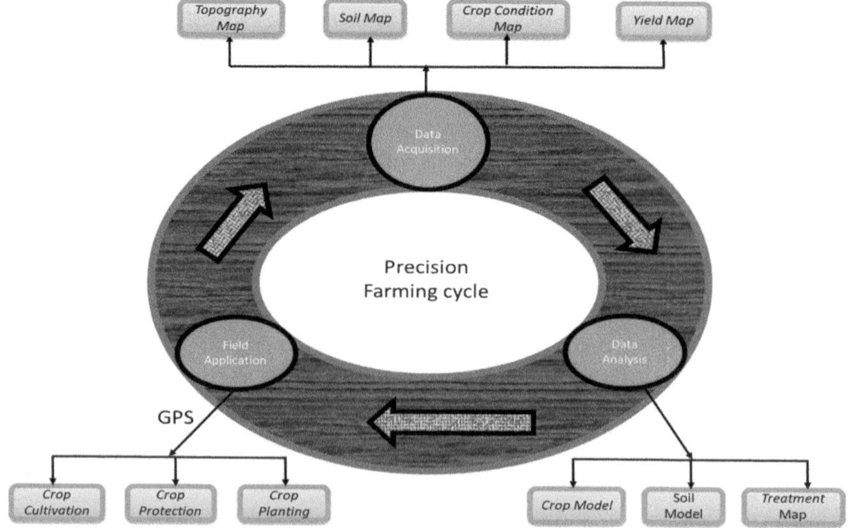


**Fig. 2.1** A systematic presentation of precision farming cycle

## 2.2 Why Precision Farming

Due to the increasing population and decreasing agricultural land, the global food production system has started facing direct and indirect challenges that will keep on increasing worldwide over the upcoming years. Therefore an immediate requirement of implementation of novel technologies, knowledge, sufficient investment, and higher agricultural funding is needed so that novel solutions to the new problems can be given to cope with the challenges. Some major issues related to the decline in agricultural development, total productivity, and growth include lack of eco-regional conservation methodology, degrading and shrinking natural resources, universal climatic dissimilarity, decreasing agricultural land, decreasing farmers' incomes, division of land possessions, restricted work opportunity in non-farming sectors, and agricultural trade liberalization (Fountas et al. 2005). Hence, the adoption of recently emerged technologies is seen as one of the key factors to increase agricultural productivity. As an alternative implementation of PF systems, which identifies the site-specific alterations according to the farmer fields and also adjusts the actions, is much beneficial than managing total the agricultural land on the basis of several average hypothetical conditions, which does not even exist anywhere in the agricultural land (Batte and Buren 1999).

Generally, farmers are quite aware of their agricultural land variable yields. The information related to agricultural land is difficult to store due to its larger data sizes and annual swings in agricultural area arrangements. Thus, PF plays a vital role in managing the whole field data. It has the potential to collect the data, simplify it, and analyze the information automatically (Shibusawa 2002). It also allows the management in decision-making and fast implementation of the decision on farming lands (Fig. 2.2).

## 2.3 Apparatus and Instruments

PF is a farming technique that is an amalgamation of various technologies and its application in the agricultural field with the aim of improving crop productivity. Every instrument used in PF is reciprocally interrelated with others. Some main components are discussed in detail below for better understanding.

### *2.3.1 GPS (Global Positioning System)*

One of the main and crucial components of PF is consists of several sets of satellites. These sets of satellites send the radio signals in the form of waves to the receiver installed on the ground; the receivers, after receiving the signals, processed it and found out the exact (95%) topographical position (Batte and Buren 1999). These

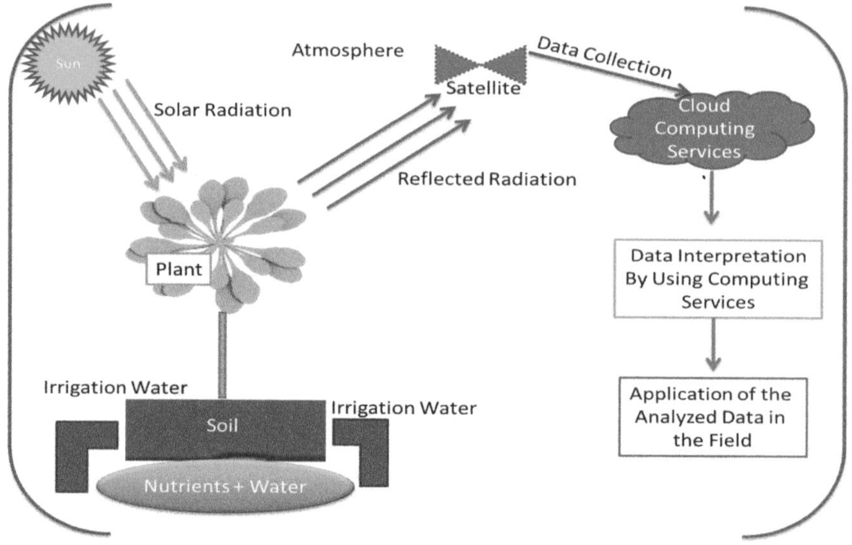


**Fig. 2.2** A diagrammatic representation of precision farming design model

systems provide the specific mapping of the farmer's agricultural land and also inform the farmers by providing the detailed description related to the crop and farm to the farmers like crop status, which area requires fertilizers, pesticides, or water by utilizing the suitable software (Lang 1992).

### *2.3.2 VRT (Variable-Rate Technology)*

The VRT technologies are the automatic system that decides the field inputs on the basis of the farm soil determined by the soil map and is the common and widely used system in the US (United States). Data obtained from the GIS system plays a vital role in managing the different processes in the farm like sowing time, selecting herbicide and weedicide, spraying pesticides or fertilizer at the specific area and time (Lang 1992).

### *2.3.3 Soil Sampling by Using Grid Technology*

In this technique of soil sampling, different samples of farm soils were collected and systematically placed in a specific grid along with the sample location detail and were examined in the research laboratories. After analyzing the sample, a conclusion is drawn for every taken soil sample regarding the crop nutrient requirement. The objective behind analyzing the soil sample by using a grid system is to create a map of crop nutrient requirements. Further, the map regarding the fertilizer need is also created by using the software (Berntsen et al., 2006).

### *2.3.4 GIS (Geographic Information System)*

The GIS system is used in PF to create a computerized map and is also used to examine the topography and other parameters, including the type of farm soil, drainage, and yield, by using a spatial and statistical approach. These GIS systems usually consist of a software (programs), hardware (equipment), along with the designed procedures in order to perform the storage, compilation, and analysis of feature aspects for generating the maps. One of the major advantages is that it links all the data at a single location, so when required, it can be easily analyzed. The GIS map that was drawn by using the computerized software contains numerous and much precise information about the different parameters like plant yield, soil information, crops detail, soil nutrient content, rainfall, and pests or weed in crop comparison with the conventional maps. Once the obtained data were analyzed, the outcomes are further used to understand the connections amongst the different features affecting the crop at a specific site (Neményi et al., 2003).

### *2.3.5 Crop and Soil Sensors*

Sensors are the integral and crucial component of PF technology. These sensors are extensively used in PF in order to obtain precise information related to soil fertility or nutrients, water status, and crop growth. For example, ECa (electrical conductivity) sensors are the most commonly and widely used sensors for characterizing the variability of farm soil, such as soil salinity and nutrients present in the soil. The main reason behind using the ECa sensors is its sensitivity to change in soil salinity and texture and also provides an outstanding reference line to implement the location-specific management (Adamchuk et al., 2004).

### *2.3.6 Sensor-Based Equipment*

Numerous technologies like conductivity, photoelectricity, sound waves, and electromagnetic are commonly used to measure moisture, temperature variation, soil texture, nutrient content of the soil, vapor pressure, etc. (Doruchowski et al., 2009). Data collected by using the remote sensing (RS) technology are specifically utilized to detect the stress situations, recognize weeds and pests in the farm, differentiate between the crop species, and monitor crop and soil conditions (Njoroge et al., 2002).

**Table 2.1** Wireless node which is commonly used in the agricultural field

| Reference | Measures | Name of the sensors |
|---|---|---|
| Aqeel-ur-Rehman and N, Islam, Z.A. Shaikh, 2014) | Plant photosynthesis | CM1000TM- Field scout |
| | Moisture Content of the Soil, Temperature of Soil, Salt Concentration | EC250- EC sensor |
| | Moisture Content of the Soil, Temperature of Soil | Pogo-Soil sensor |
| | Various Air parameters (Temperature, Humidity, and Pressure), Speed of Wind | CM-100-Weather Sensor |
| | Various Air parameters (Temperature, Humidity, and Pressure), Speed of Wind | MSO-Met Station One |
| | Temperature and Humidity | SHT75, SHT71-Sensors |
| | Various Air parameters (Pressure, Humidity, and Temperature) | HMP45C-H-chip Sensors |
| | Various Plant parameters (Wetness, Temperature, Moisture) | 237 leaf-Sensor |
| | Plant Photosynthesis | YSI 6025-Sensor |
| | Crop Temperature | Leaf Temperature-2 M sensor |
| | Various Air parameters like Humidity and Temperature | Cl-340- Sensors |

Various sensors used in the PF allow the assemblage of enormous amounts of information without laboratory investigation (Chen et al., 2004). Some common sensors which are used in the agricultural field are further described in Table 2.1.

## 2.3.7 *Monitoring of Crop Yield*

Monitoring of crop yield typically consists of various sensors along with the information storage device, a display (monitor), along with the computer (loaded with the set of programs) positioned in a systematic way in a single location, which manages the communication and amalgamation of these components. These sensors will calculate the grain (the volume or mass) flow, speed of the separator, and ground speed. The grain yield is determined by calculating the grain flow force as it touches the grain elevator sensible plate. Several newly designed mass flow sensors working are based upon the principle of beams transmitted and calculating the energy portion that comes back after striking the flowing seeds. In the case of grains, yield is continuously recorded by measuring the force of the grain flow as it impacts a sensible plate in the clean grain elevator of the combine. The recent development of a mass flow sensor works on the GPS system receivers are utilize for the purpose of creating crop yield maps and recording the yield-related information position. On the other side, these yield monitoring technology consists of devices to keep track of grains moisture, and weight at the specific location (Davis et al., 1998).

## 2.4 Image-Based Sensing

For some decades, remote sensing (RS) technology is extensively utilized in PF to analyze the crops' conditions. In this technology, the Earth's physical situations were remotely noticed by measuring the emitted radiation and reflected radiation. High-resolution (magnification) cameras are used for the purpose of capturing the images for further investigation. Several supporting platforms are used for these cameras.

### *2.4.1 Satellite Platforms*

Among all the used platforms, a satellite sensors platform is considered as one of the most stable platforms for RS (remote sensing), comprises of rockets, space shuttles, satellites, and are characterized on the basis of timing and orbits (Toth and Jóźków, 2016). The main benefits of these satellite platforms are their high spatial resolution, stable and clear image, cover a huge area, and noise-free images. Apart from its advantage, it also has some disadvantages also such as the high cost for the images of high spatial resolution, cloud will affect the land features (the satellites are weather sensitive) in the captured image, and it has a fixed schedule (Rudd et al., 2017).

### *2.4.2 UAV (Unmanned Aerial Vehicle) Platforms*

This platform has several advantages, including its low cost, flexible accessibility, replaceable sensors, and ability to capture images of very high spatial resolution. Because of these advantages, it can be used as an alternative to airborne and satellite systems. Usually, this UAV platform comprises of navigation and communication system along with a set of sensors, and its flying time depends upon the weight of the load (Zhong et al., 2018). More detailed information related to the crop can be concluded from the images captured via UAV platform than the images captured via satellite platforms (Candiago et al., 2015). Further, in order to assess the crop conditions, three different types of VI (Vegetation Index) maps are used.

### 2.4.3 Airborne Platforms

Compared with the other platform, i.e., satellite sensor platforms, airborne sensor platforms are quite flexible but costly. The main advantage of this type of platform (airborne) over satellite sensor platform is its ability to capture crop images of very high spatial resolution. On the other hand, its coverage area is considerably lesser than the satellite sensor platforms; however, its coverage area is relatively larger in comparison with the UAV sensor platforms. Several commonly used airborne sensor platforms are helicopters, gas balloons, airplanes, and aircraft (Rudd et al., 2017).

## 2.5 Application of WSN in Agriculture

WSN (wireless sensor-based networks) has numerous applications in the growing agriculture sector. Several common applications of WSN in the agricultural field are smart (irrigation, pest control, and fertilization) areas.

### 2.5.1 Advanced Pest Management Technology and Early Disease Prediction Methodology

In the agriculture sector, the main reason behind the low productivity of crops is pest attack. These insect pests cause severe damage to plants, also responsible for impeding growth development in crop plants. However, forecast before the onset of the disease is generally useful for the farmers to make precise decisions for controlling the disease before it causes too much loss. Pest management systems consist of different types of electronic equipment that facilitate humans to traps insects-pest within a particular range of this electronic gadget (Mahlein et al., 2012). These instruments are composed of sensors that have the capacity to predict environmental fluctuations for future analysis. Many studies have been conducted in the field of agriculture for pre-disease diagnosis and pest management systems using highly developed techniques (Mahlein, 2016). There are several types of sensors like Fluorescence imagery sensors, spectral sensors, Thermal sensors, RGB sensors applied by various researchers to record imagery data (Lee et al., 2017). The thermal sensors are applied for accurate measurement of water level in the crops by recording the ambient temperature since it is directly proportional to water content in plants. RGB sensors provide a biometric effect in crop plants which is perceived with the help of three color channels, i.e., red, green, and blue. Geographical and object images are captured by Multi and hyperspectral sensors having multiple wavebands. The high image resolution is directly proportional to the distance between the object and the sensor. In a similar way, drones have high spatial resolution due to their lower altitude as compare to satellite images. The rate of photosynthesis in the plant is measured by fluorescence sensors activities in the plants. To diagnose various diseases in plants, different image processing methods have been applied for virtual data analysis. In a given study, an IoT system approach was utilized for pest and plant disease prediction so to reduce the excessive application of different fungicides and pesticides (Lee et al., 2017). Different aspects of weather like rain, dew, temperature, sunshine, wind, velocity, wind direction, etc., are monitored by the sensor in the correlation between pest density and plant growth. The sensors applied in the field record the data and store it in the cloud. Before the onset of alarming situations of serious pest attacks on crop plants, farmers are being informed. Generally, hyperspectral images are utilized to evaluate crops' health and insect attacks by deploying spectral cameras. The images captured by cameras are evaluated using a machine learning process to finally diagnose the disease in crops. Advance Neural Networks (ANNs) are commonly applied for analyzing imagery data because of their capability to read complex structures and different kinds of patterns. Hyper-spectral imaging methodology was applied to detect diseases and insect-pest attacks in crops (Golhani et al., 2018). An ANN having multiple layers showed high performance for disease diagnostics. A similar line of work disease identification in sugar beet crops was observed at a very early stage using this technology (Rumpf et al., 2010). For early identification of disease, four different types of algorithms were considered on spectral images. Spectral images were then collected and analyzed, and multiple vegetation indexes were recorded, which then provided data for prospective analysis. The vegetation indices involved in the study include SIPI, NVDI, SR, PSSRa and PSSRb, ARI, REP, mCAI, and RRE. These vegetation indices data were stored in the dataset. SVM, Advanced Neural Network, and decision tree were further utilized for classifying the data type. In a comparative study, the SVM classifier resulted in a precise diagnostic of disease with the precision of approximately 97.12% as compared to another classifier. Further data mining approach was applied to the already collected information for two different crops, i.e., wheat and rice, in India. For reduction in dimension, Sammon's mapping technique for scaling was applied to decrease the dimension (Sanghvi, 2015). For high-dimensional data, dimension reduction is a necessary requirement that aids in better visualization of data with accurate prediction. It provides better results and is effective for any algorithm. Principle component analysis is one the important technique which is simultaneously applied with Sammon's mapping for better results. In this technique, multidimensional data is reduced to two or limited dimensions. Then, the Self-Organized Mapping (SOM) algorithm was applied to describe the correlation among the data based on clustering. The comparison between both techniques SOM and Sammon's mapping suggested that SOM was well suited for a large dataset; on the other hand, Sammon's mapping provided better results for a small subset of data. Smartphones were also involved in the process of data acquisition, which was utilized for assistance in crop growth and developmental process and played a crucial role in the prediction of crop health by analyzing the datasets (Hufkens et al., 2019). Wheat crop health was studied with the help of the images captured by high-resolution mobile cameras. The wheat crop was then grouped under two categories, healthy and unhealthy, supported by green level by computing Gcc technique. Generally, most of the techniques in PF involve the use of multiple sensors that are IoT-based, which is utilized to assess the crop health, or remote sensing approach, which is based on analyzing spectral image data utilizing various algorithms. Monitoring of crop health is possible based on some of the traits in which sensors are applied for a particular purpose, despite of unavailability of web or mobile services. To obtain precise information about crop growth and development, both IoT approach and remote sensing technology should be utilized together for proper results. In the agriculture sector, web and mobile facilities could be utilized for the prediction of crop health based on imagery data.

### 2.5.2 *Precise Irrigation Approach by Using Advanced Computational Techniques*

Smart irrigation refers to the system of irrigation in which time, the water requirement of the crop are the main parameters that are taken into consideration while watering the crops. It is one of the major aspects of agriculture, which is deciding factor for crops' health and productivity. Smart irrigation aims to control the wastage of water and its judicious utilization for agriculture purposes since most of the nation across the globe face serious water shortage. An example of this irrigation system was practiced in Raspberry Pi (Akubattin et al., 2016). Two sensors were used: The first—the soil moisture sensor, and the second, temperature and humidity sensor. The first one was utilized to measure soil water level, while the second one was involved in monitoring the environmental fluctuations. The Raspberry Pi was connected with an artificial irrigation system having sensors.

A mobile function to channelize and control the flow of water was developed, enabling both manual and automatic control. In an automatic system, water flow was controlled automatically, which was supported by the water level in soil without the involvement of humans. In manual mode, human intervention was needed to check soil moisture. An alarm was provided if the water level fell below the threshold level. The user turned ON the mobile application to maintain desired soil moisture. The continuous power supply is a major area of concern in IoT-dependent technology, so many scientists have focused their research on the development of power-efficient systems. A water irrigation system that utilizes solar energy is a type of power-efficient system (Harishankar et al., 2014). In soil sensor and water supply were connected to a specific controller system. The water supply was further controlled by a valve that could be was turned ON/OFF based on water requirement as indicated by the soil moisture sensor. The energy for the purpose of irrigation was provided with the help of a solar panel, so the system doesn't require an external source of energy. Another power-efficient system that utilized sensor-supported by IoT technology was controlled by opening and closing a solenoid valve based on the soil moisture requirement. In addition to this, weather alerts signal were sent to the farmers through mobile services to update the present environmental parameters like temperature and relative humidity, which directly had affect soil moisture content. In a power-efficient method of irrigation, crop cultivation requires wireless networking sensors that effectively utilize water in different environmental conditions (Nikolidakis et al., 2015). This irrigation technique is based on various factors like humidity, wind velocity, wind direction, and temperature, and the data recorded by sensors based on these climatic conditions.

In an IoT-based technology, the soil water level is controlled by ATMEGA 328P with a General Packet Radio Service system (Rawal, 2017). The data obtained by these sensors were sent to cloud computing, i.e., where graphs and PCA are generated to analyze the pattern and trend followed by datasets. An internet-based portal was also created for the farmer where the water level can be continuously monitored. A better-advanced method of irrigation was developed in a similar line of work to measure soil water level by soil moisture sensor and soil temperature sensors. RFID was applied to transfer data to the cloud for the purpose of data analysis. Using an advanced water sprinkler, irrigation was done in various crops by taking various environmental parameters like temperature, humidity, and soil water level with the help of sensors (Kumar et al., 2017). The water content from the sprinkler was channelized based on the soil moisture depletion and further to reduce human intervention. The replacement of sensors is a major concern that affects the accuracy of data recorded. A detailed description of soil moisture sensor is provided in this chapter that is useful in deterring water requirement under severe drought conditions so to use water judiciously in the water-deficient area (Soulis et al., 2015).

### 2.5.3 *Morden Fertilization Approach Based on IoT and Sensors Techniques*

Fertilizers are chemical substances that are applied in crop fields for high productivity and yield. They are basically inorganic and organic in nature that is responsible for providing nutrients to plant for proper growth and development. The manual process of spraying nutrients in solution form by mixing fertilizer with water and dusting is a normal technique applied by farmers for fertilization. However, the advanced procedure of fertilizer application includes knowing what quantity of fertilizer is required by the plant. It is significantly important to apply fertilizer to plant at the right time and inappropriate amount to increase the yield and productivity of crops (Cugati et al., 2003). Various fertilizer application techniques have been evolved and described by researchers across the globe since the last few decades using WSN, and IoT approaches. A recent advanced fertilization system automatic in nature was applied, which involved the use of real-time sensors to determine soil fertility and productivity (He et al., 2011). This system takes into consideration three modules which include input, output, and decision-making system. The real-time sensor and the data generated by it are the main component in the decision support module to estimate the accurate amount of fertilizer required by the plant for its growth and development. An advanced sensor system of mechanical nature named "Pendulum Meter" was deployed to measure the optimal fertilization dose required by the plant. This sensor was firmly fixed on the top of the tractor to estimate the density of different crop plants.

Further, the fertilizer spreader that was attached with the mechanical pendulum meter recorded the data generated by this sensor (Chen and Zhang, 2006). The IEEE 802.11 Wi-Fi module, along with GPS, was utilized for the data transfer process. Different types of sensors which include—soil moisture content, temperature, NO2, CO2, etc., include real-time data and utilized GIS to analyze that information generated by sensors.

## 2.6 Major Problems of Precision Farming

For the last few years, PF has been generally used to enhance the yield of various crops, with decreased human effort and costs, even though the farmers' adoption of these techniques is still very less in number due to the following problems or reasons. The major problems related to PF are mention in Fig. 2.3.

### 2.6.1 Data Management

The sensors system used in PF regularly generates data. Thus the generated readings should be in sequence so that the appropriate decision should be taken timely when required, and false decision-taking situations should be avoided. The PF systems generate enormous amounts of readings, and it requires adequate resources to do the

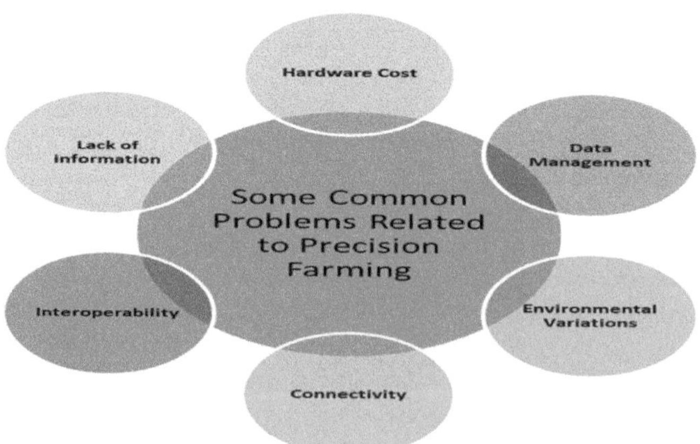


**Fig. 2.3** Some common problems related to PF

analysis of the generated data (Robert, 2002). Due to which new software facilities and platforms for the analysis of such a huge data is demand and is required.

### *2.6.2 Environmental Variations*

Changing weather is among one of the main problems which directly disturbs the sensors' accuracy of collecting the data. Especially the wireless nodes sensor installed in the farming land is very sensitive to various changing weather conditions including temperature fluctuation, sunlight, changing wind speed, rain, etc., the components such as drone and satellite are also affected by the atmospheric variations. Signal communication between the cloud and sensor nodes can be disturbed due to the interference caused by atmospheric disruption. The photos captured are also very much affected by the various contaminants present in the atmosphere (Robert, 2002). Still, the development of improved systems so that these types of problems can be solved is a challenge for the researchers.

### *2.6.3 Accessibility of Farm Insight Information*

Outcomes obtained from the agricultural land by using farm-based WSN will instantly become available at the farm, while further delays, which may be due to external communication, can be avoided. This immediate availability of the information is mainly important for the agricultural fields located in areas where the Internet connectivity is very poor (Akridge and Whipker, 1999).

### *2.6.4 Hardware Cost*

PF mainly depends on the hardware, including sensors, small sized-drones, wireless nodes, which help in assessing the various factors in real-time. But the used sensors have various drawbacks such as high development price, high repairs price as well as high positioning cost. On the other hand, several PF systems which are fit for small land are of low cost, like smart irrigation systems, requiring low-cost sensors as well as the hardware components. Conversely, drone systems which are used to monitor the crops growth are only feasible for large agricultural land because of their high installation and maintenance cost (Akridge and Whipker, 1999).

### 2.6.5 Data Security

Analyzing several raw data analytics at the same place (within the farm) also has numerous advantages from the data security point of view. Hence, only the summary of field data, like only the average of specific parameters across the agricultural land, will only require to be exchanged with the third party, i.e., cloud analytics. Thus, the data shared with the third party becomes quite complicated, and it becomes difficult for them to obtain the sensitive information related to the farm from the shared information (Pierce and Nowak, 1999).

### 2.6.6 Exchange of Information

Apart from the above-discussed problems, interoperability or data exchange is among one of the biggest problems faced in PF. The lack of data sharing not only reducing down the growth but also blocking the adoption of new technologies such as IoT and also obstructs the improvement in crop production efficiency obtained by using smart agricultural systems (Pierce and Nowak, 1999). Thus in these conditions, some new or modified protocols and methods need to incorporate with the machine communication systems to increase the system-to-system communication efficiency, and data exchange between the management systems and machines is required.

### 2.6.7 System (Computing) Sustainability

Eventually used for the purpose of enhancing the life of sensors battery, power harvesting is an integral part of the latest WSN design. Technically operation of WSN is normally powered (partially) by using different renewable energy sources, like wind and solar energy. Thus, various analytical work performed within the system will also receive power from renewable sources, which results in more environment-friendly and a greener way of computing (Pedersen et al., 2001).

## 2.7 Conclusion and Future Prospects

PF is a new or latest farming practice used by farmers to enhance the productivity of crops by utilizing modern technologies such as IoT (internet of things), AI (Artificial Intelligence), WSN, ML (Machine Learning), and CC (cloud computing). Till date, most research studies conducted related to PF so far show that PF-based farming has greatly influenced crop productivity and sustainability. The objective of precision farming is to support the farmers in making a decision on the basis of multiple analyzed parameters of agricultural crops, like soil water level, sunlight intensity, soil nutrients, temperature, speed of the wind, and humidity. The key focus of PF is to increase the yield of cultivated crops by optimizing different available resources like pesticides, water, fertilizers, temperature, etc. On the other side, prescription maps (PM) play an essential part in resource optimization by enabling the farmers to identify the resources which are required at specific growth time for the production of healthy crops. But still, there are some problems are faced in the development phase and installation phase of these PF systems. This chapter provides a review of the latest technologies used in PF, their applications in smart agricultural systems, and some common problems related to PF. Maximum research related to PF (agriculture domain) is based on the RS (remote sensing) platforms to collect the images, which only reflects the VIs (Vegetation Indices). While generating the PM (prescription maps), various other related factors should also be carefully taken into consideration, like moisture level of the soil, weather-related parameters, etc.

## References

Abbasi AZ, Islam N, Shaikh ZA (2014) A review of wireless sensors and networks' applications in agriculture. Comput Stand Interfaces 36:263–270

Adamchuk VI, Hummel JW, Morgan MT, Upadhyaya SK (2004) On-the-go soil sensors for precision agriculture. Comput Electron Agric 44:71–91

Akridge JT, Whipker LD (1999) Precision agricultural services and enhanced seed dealership survey results. Center for Agricultural Business, Purdue University. In: Staff Paper No. 99–6

Akubattin V, Bansode A, Ambre T, Kachroo A, SaiPrasad P (2016) Smart irrigation system. Int J Sci Res Sci Technol 2:343–345

Rehman A, Abbasi AZ, Islam N, Shaikh ZA (2014) A review of wireless sensors and networks' applications in agriculture. Comput Stand Interfaces 36:263–270

Batte MT, Van Buren FN (1999) Precision farming—factor influencing productivity. In: County O (eds) Translator Northern Ohio crops day meeting, Wood, January 21

Berntsen J, Thomsen A, Schelde K, Hansen OM, Knudsen L, Broge N, Hougaard H, Hørfarter R (2006) Algorithms for sensor-based redistribution of nitrogen fertilizer in winter wheat. Precis Agric 7:65–83

Candiago S, Remondino F, De Giglio M, Dubdini M, Gattelli M (2015) Evaluating multispectral images and vegetation indices for precision farming applications from UAV images. Remote Sens 7:4026–4047

Chen X, Zhang F (2006) The establishment of fertilization technology index system based on "3414" fertilizer experiment. China Agric Technol Ext 22:36–39

Chen F, Kissel DE, West LT, Adkins W, Clark R, Rickman D, Luvall JC (2004) Field scale mapping of surface soil clay concentration. Precis Agric 5:7–26

Cugati S, Miller W, Schueller J (2003) Automation concepts for the variable rate fertilizer applicator for tree farming. In: Proceedings of the 4th European Conference on Precision Agriculture. Berlin, Germany, June 15–19, pp 14–19

Davis G, Casady W, Massey R (1998) Precision agriculture: an introduction

Doruchowski G, Balsari P, Zande JC (2009) Precise spray application in fruit growing according to crop health status, target characteristics and environmental circumstances. In: Proceedings of the 8th Fruit, Nutritio and Vegetable Production Engineering Symposium. Concepcion, Chile, pp 494–502

Fountas S, Blackmore S, Ess D, Hawkins S, Blumhoff G, Lowenberg-Deboer J, Sorensen CG (2005) Farmer experience with precision agriculture in Denmark and US Eastern Corn Belt. Precis Agric 6:121–141

Golhani K, Balasundram SK, Vadamalai G, Pradhan B (2018) A review of neural networks in plant disease detection using hyperspectral data. Inf Process Agric 5:354–371

Harishankar S, Kumar RS, Sudharsan KP, Vignesh U, Viveknath T (2014) Solar powered smart irrigation system. Adv Electr Comput Eng 4:341–346

He J, Wang J, He D, Dong J, Wang Y (2011) The design and implementation of an integrated optimal fertilization decision support system. Math Comput Modell 54:1167–1174

Hufkens K, Melaas EK, Mann ML, Foster T, Ceballos F, Robles M, Kramer B (2019) Monitoring crop phenology using a smartphone based near-surface remote sensing approach. Agric Meteorol 265:327–337

Kumar BD, Srivastava P, Agrawal R, Tiwari V (2017) Microcontroller based automatic plant Irrigation system. Int Res J Eng Tenchnol 4:1436–1439

Lang L (1992) GPS, GIS, Remote sensing: an overview. Earth Obs Mag 23–26

Lee H, Moon A, Moon K, Lee Y (2017) Disease and pest prediction IoT system in orchard: a preliminary study. In: Proceedings of the 2017 Ninth International Conference on Ubiquitous and Future Networks (ICUFN). Milan, Italy, July 4–7, pp 525–527

Mahlein AK (2016) Plant disease detection by imaging sensors-parallels and specific demands for precision agriculture and plant phenotyping. Plant Dis 100:241–251

Mahlein AK, Oerke EC, Steiner U, Dehne HW (2012) Recent advances in sensing plant diseases for precision crop protection. Eur J Plant Pathol 133:197–209

Mumtaz R, Baig S, Fatima I (2017) Analysis of meteorological variations on wheat yield and its estimation using remotely sensed data. A case study of selected districts of Punjab Province, Pakistan (2001–14). Ital J Agron 12

Neményi M, Mesterházi PA, Pecze Z, Stépán Z (2003) The role of GIS and GPS in precision farming. Comput Electron Agric 40:45–55

Nikolidakis SA, Kandris D, Vergados DD, Douligeris C (2015) Energy efficient automated control of irrigation in agriculture by using wireless sensor networks. Comput Electron Agric 113:154–163

Njoroge JB, Ninomiya K, Kondo N (2002) Automated fruit grading system using image processing. In: Proceedings of the 41st SICE Annual Conference. pp 1346–1351

Pedersen SM, Ferguson RB, Lark RM (2001) A multinational survey of precision farming early adopters. Farm Manag 11:147–155

Pierce FJ, Nowak P (1999) Aspects of precision agriculture. In: Sparks DL (ed) Advances in agriculture. Academic Press, Cambridge, pp 1–85

Rawal S (2017) IOT based smart irrigation system. Int J Comput Appl 159:880–886

Robert PC (2002) Precision agriculture: a challenge for crop nutrition management. Plant Soil 247:143–149

Rudd JD, Roberson GT, Classen JJ (2017) Application of satellite, unmanned aircraft system, and ground-based sensor data for precision agriculture: a review. In: Proceedings of the 2017 ASABE Annual International Meeting, Spokane, Washington, United States, July 16–19

Rumpf T, Mahlein AK, Steiner U, Oerke EC, Dehne HW, Plümer L (2010) Early detection and classification of plant diseases with support vector machines based on hyperspectral reflectance. Comput Electron Agric 74:91–99

Sanghvi Y et al (2015) Comparison of self organizing maps and St Ammon's mapping on agricultural datasets for precision agriculture. In: Proceedings of the 2015 International Conference on Innovations in Information, Embedded and Communication Systems (ICIIECS), Coimbatore, India, vol 54, March 19–20:1–5

Shibusawa S (2002) Precision farming approaches to small farm agriculture. Agro Chem Rep 2:13–20

Soulis KX, Elmaloglou S, Dercas N (2015) Investigating the effects of soil moisture sensors positioning and accuracy on soil moisture based drip irrigation scheduling systems. Agric Water Manag 148:258–268

Toth C, Jóźków G (2016) Remote sensing platforms and sensors: a survey. ISPRS J Photogramm 115:22–36

Zhong Y, Wang X, Xu Y, Wang S, Jia T, Hu X, Zhao J, Wei L, Zhang L (2018) Hyperspectral remote sensing: from observation and processing to applications. IEEE Geosci Remote Sens Mag. 6:46–62

# Chapter 3
# Smart Dairy Farming Overview: Innovation, Algorithms and Challenges

**Sindiso M. Nleya and Siqabukile Ndlovu**

**Abstract** The recent increase in world population has correspondingly triggered an increase in dairy demand subsequently giving rise to the global hunger problem. The global hunger problem translates to some sections of the world population facing chronic food deprivation and malnutrition in the form of nutrient and micronutrient deficiencies as well as shortfalls in vitamins and essential metals. In mitigating the global hunger problem, the Smart Dairy Farming paradigm is leveraging sensors, Internet of Things (IoT), broadband technologies and data analytics to craft innovative solutions and systems. These innovations are crafting systems, which are derived from applying Machine Learning (ML) algorithms on the big data generated from the numerous sensors and IoT equipment in the dairy farm. Specifically, the innovative solutions are aimed at not only improving milk yields but also enhancing the efficiency of the dairy process. At the product level, the innovative systems strive to increase milk production by deploying robotic milking systems that milk the cow, analyse the milk, process the milk and preserve it. At the process level, the systems are concerned with the health and welfare of the very cow producing the milk as they can monitor cow movement, feed and health.

**Keywords** Smart Dairy · Analytics · Process · Innovation · Yield

## 3.1 Introduction

Due to the increasing world population, milk consumption is continuously increasing (Taneja et al. 2019). Statistics show that dairy products' consumption (milk, cheese, cream) is generally higher in developed countries (see Fig. 3.1). According to OECD and Food and Agriculture Organization of the United Nations (2020), the global

S. M. Nleya (✉) · S. Ndlovu
Computer Science Department, National University of Science & Technology, Box AC 939, Ascot, Bulawayo, Zimbabwe
e-mail: sindiso.nleya@nust.ac.zw

S. Ndlovu
e-mail: siqabukile.ndlovu@nust.ac.zw

 
A. Choudhury et al. (eds.), *Smart Agriculture Automation using Advanced Technologies*, Transactions on Computer Systems and Networks,
https://doi.org/10.1007/978-981-16-6124-2_3

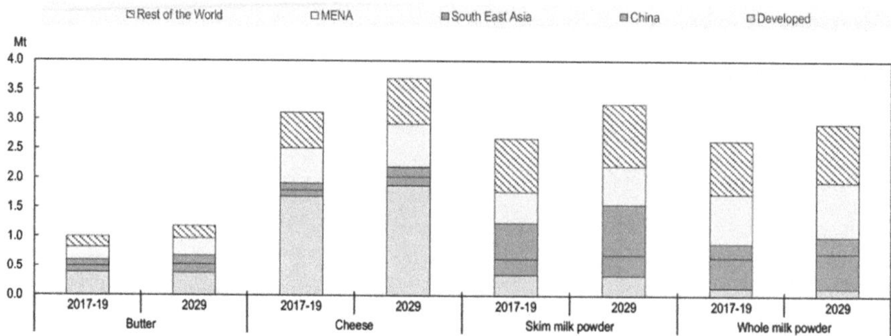


**Fig. 3.1** Milk consumption in selected countries and regions (OECD and Food and Agriculture Organization of the United Nations 2020)

consumption of dairy products will increase over the coming decade due to income and population growth. World per capita consumption of fresh dairy products is expected to spike by 1.0% per annum over the next 10 years. This is a faster increase compared with the previous decade. With this rising demand, it is important for farmers to increase their yield. Dairy farmers, however, face several challenges to run and maintain their businesses. These challenges include climate changes, herd management, labour shortages, etc. With these constraints, dairy farmers need to concentrate on two things: monitoring the health of their animals (to detect health issues early) and improving reproduction and milk yield (so that they have more cows) to increase productivity and profits (Taneja et al. 2019).

Better technological techniques like the Internet of Things (IoT) and data-driven approaches are required for improving milk production (Akbar et al. 2020). The IoT is changing the way we live and make decisions. Situations demanding instantaneous processing and decision support have been on the increase (Vasisht et al. 2017). IoT can be applied in dairy farming to help with real-time processing of data to improve decision-making. Clearly, for the effective IoT functionality, broadband is also important as it extends coverage to rural and remote areas using the approaches (Brown et al. 2014; Nleya 2016) at lower costs.

The application of IoT in dairy farming is referred to as smart dairy farming (SDF) or precision dairy farming.

## 3.2 Smart Dairy Farming

Smart dairy farming integrates the IoT, precision farming techniques, big data analytics and cloud computing to increase productivity in the dairy industry (Kulatunga et al. 2017). Smart dairy farming uses IoT and precision farming techniques to collect data around a dairy farm and use this data (big data) to improve milk production. For example, milk yield and milk quality are rarely associated with other variables such as animal health, weather and feeding. However, the availability of these data from Internet-connected farm devices allows farmers to make more timely and informed decisions (Lokhorst 2018). Smart dairy farming can also reduce environmental issues, cut down on resources and improve animal health (Akbar et al. 2020). It is, therefore, necessary for a well-managed farm to take advantage of IoT technologies to manage and control all the activities of the farm from management to the actual production of milk.

Examples of technologies used in smart farming include milking robots, automatic feeders and sensor systems used to manage individual animals. Smart dairy farming uses data to sustain and improve dairy farming. These data, which are collected from the farm, provide the capacity to gain a deep understanding of the production of milk and the overall health of the animals. In line with this, accurate decisions are drawn up that are intended to extend the life expectancy of a cow and to optimise milk production (Newton et al. 2020).

### 3.2.1 Functions of Technology in Dairy Production

It is always hard to correctly determine if a cow is coming down with disease, the right time for milking, the correct amount of feed to give, etc. These factors in a dairy farm eventually affect the quality of the milk produced. Technology can support farmers in this regard. The use of IoT devices (on the farm and in/on the animals) like sensors (wearable/not wearable) can inform the farmers of the status of every cow at any given time. A sensor is defined as an electronic device that measures a certain parameter of an individual animal and automatically detects changes in the parameter being measured that is related to the health of the animal and requires the farmer to take action (Akbar et al. 2020). The use of sensors can effectively detect: the illness of individual animals (before that illness affects the other cows or even the milk production), temperature anomalies in the sheds, weight loss of an animal, reduction in milk production, poor quality of milk, etc. (Akbar et al. 2020).

Technology has had a great and positive impact on dairy farming processes. It has helped contribute to making dairy farming efficient and easier resulting in lowered costs and improved yields (Wolfert et al. 2017). The use of technology in farming has increased the variety of dairy products in the market. Technology in dairy farming is used for a number of processes like: milking, recording the yields of milk, monitoring milk quality (such as the components of protein, fat and other minerals), recording the activity and movement of individual animals, recording temperature, detecting fertility and disease and taking measurements of the animals' body weights. SDF is made up of four major elements: management, environmental control, single animal information and automation of rearing work. Figure 3.2 shows the elements of SDF as proposed by the Smart Dairy Consortium (Lokhorst 2018).

As indicated in Fig. 3.2, SDF is made up of a number of elements. These elements show that the need for technological support on dairy farms is gaining importance. Important issues must be immediately solved, such as monitoring the health and

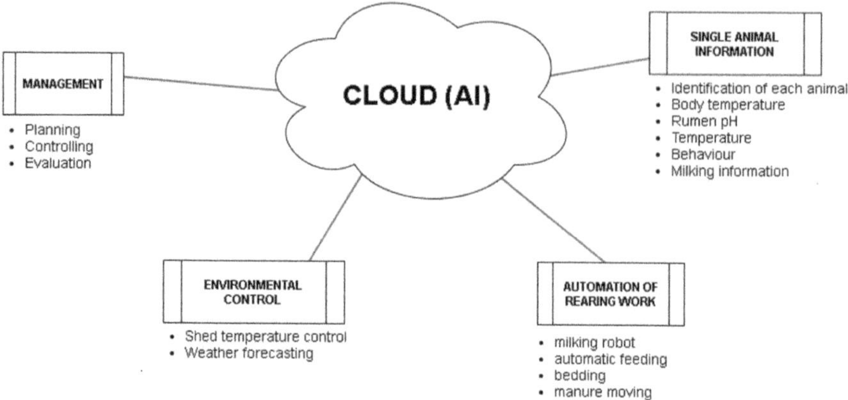


**Fig. 3.2** Elements of smart dairy farming

welfare of individual animals, reducing the impact farming has on the environment, and assuring productivity (Berckmans 2017). The use of advanced technologies to manage and monitor each animal in a farm is called precision livestock farming (PLF). The pioneering massive application of PLF was the electronic milk metre for cows used in dairy farms (Berckmans 2004). The milk metres were commercially used in the 1970s (Peles 1978) and early 1980s.

In the next section, we are going to discuss the elements of a smart dairy farm using Fig. 3.2 as a reference.

### 3.2.2 *Elements of a Smart Dairy Farm*

(i) Environmental Control

In SDF, digital technologies provide reliable continuous monitoring and measurement of the physical environment. The physical environment of concern here is the livestock buildings (cow sheds), the farm and the environment around the farm. IoT tools like sensors and cameras can be used to monitor the environment. It is important for dairy farmers to control the environment of cow sheds to keep the animals happy and comfortable. Controlling the farm and the surrounding environment helps reduce the environmental impact caused by dairy farming processes.

Climate can be identified as one of the main detrimental factors of production in a dairy farm (Fournel et al. 2017). Too much heat in the cow shed causes thermal stress, which results in morbidity, and mortality of animals. Precision livestock farming technologies enable the automatic monitoring of the environment and behaviour of animals (Lovarelli et al. 2020). Devices like temperature sensors, gas sensors, humidity sensors and cameras can be used here. The data collected from these technologies can be merged and used to continually assess the livelihood of the animals in relation to their environment.

IoT allows for the continuous monitoring, modelling and control of animal bio-responses, which optimises animal performance. The dairy industry is one of the most controlled industries in farming (Tullo et al. 2019). Farmers intervene to avoid pollution in the land, water and air around them. Precision livestock farming, compared with traditional livestock farming, reduces the environmental impact of the effects of farming (Guarino et al. 2017). PLF aims to make livestock farming sustainable (economically, socially and environmentally) through the use of technology to observe, interpret and control the activities of the farm. When PLF is used to reinforce management policies, it may cause a reduction of the environmental impact of farms (Quan et al. 2021). A number of studies are reporting PLF as efficient in reducing the environmental impact.

According to Tullo et al. (2019), the introduction of PLF in farms can reduce the emission of ammonia and greenhouse gases in the air, pollution of water bodies by nitrates and antibiotics, and the release of phosphorus, antibiotics and metallic chemical elements in the soil. They encourage further investigation into the positive effects of PLF on the air, soil and water. Choi et al. (2021) constructed a sustainable process to take care of livestock waste in order to reduce the environmental impact of farming. In their study, they enhanced the value of livestock waste into more valuable gases, biochar and syngas. Syngas is synthetic gas that is basically hydrogen and carbon monoxide. It is used to manufacture other important gasses and alcohols like hydrogen, methanol and ethanol (Yang and Ge 2016). Biochar has a remarkably environmentally friendly effect of reducing emissions from greenhouses and increasing soil fertility (Ashiq and Vithanage 2020). Biochar can also be used to treat contaminated soils by deactivating contaminants thereby reducing their environmental risks to humans (Shen et al. 2020).

(ii) Management

Dairy farm management is similar to other management responsibilities with three key management functions as shown in Fig. 3.3. These functions are

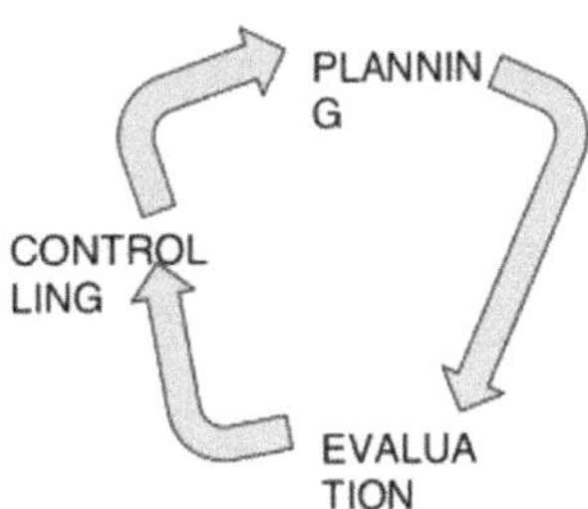


**Fig. 3.3** Management functions

overlapping in nature as each function blends into the other, and each affects the performance of the others.

Managing a dairy farm involves decision-making on different levels. It is important to distinguish between the different levels of management (Lokhurst 2018).

Low-level management, also known as operational management, focuses on executing specific tasks such as temperature control, feeding, animal monitoring or milking. Sensors and timers can be used at this level of management to detect the temperature of the shed, the level of water and feed. Sensors can also be used to monitor the wellness of each individual animal. Examples of decisions and actions at that level include turning on the aircon and opening ventilators if the temperature in the shed is too high, giving a specific cow more food, inseminating the cow that is in heat, or trimming a hoof because a particular cow shows severe movement problems.

Middle level or tactical management executes approved plans on a specific process within a farm. Sensors can be used here to detect the amount of milk in a cow so as to determine when it can be ready for milking. Sensors and other devices can be used to assess the quality of the milk so that only the best milk is produced. An example of an activity at this level is making a milking plan. Planning for immunisations is also a good example.

High-level or strategic management considers decisions that affect the whole farm. High-level management decisions have long-term repercussions and sometimes need a lot of financing. The farmers can track disease in real time and keep a record for future analysis. This maintains the milk quality and suggests requirements for healthy milk production. Examples of activities in this level are: the decision to build a new cow shed or buying a bigger milking machine.

(iii) Single animal information

Sensors are being used to provide information for decision support systems used in smart dairy farming. Sensors can be attached (fitted on the inside or the outside of the body) or non-attached (not fitted). Sensors provide actionable feedback from the animals being monitored. This is superior to the traditional model of farming, which is based on decision-making of human observations and the experience of the farmer. Figure 3.4 highlights some of the sensors used in SDF. Each of the sensors collects data about a variable on an individual cow and that data are fed into a system and collated to observe the animals. Analysis of the data collected from the sensors can then be used to provide farmers with real-time information to help diagnose risks and predict future consequences (Lioutas et al. 2019).

(iv) Sensors on the animal

The electric ear tag can be used for cow location tracking (Zin et al. 2020) or monitor body temperature and food intake (Long et al. 2020). If used to monitor body temperature and food intake, an electric ear tag can help detect a sick animal. Another IoT device that can be used to measure temperature

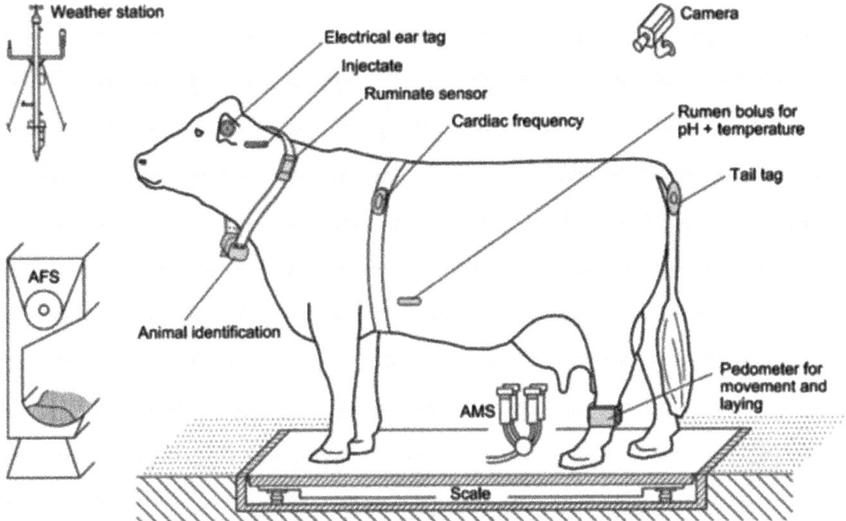


**Fig. 3.4** Sensors on a dairy cow (Bauerdick et al. 2019)

is the thermal infrared sensor. It uses infrared radiation levels to measure the temperature of an individual cow.

Pedometers are used to measure the total number of steps that each animal takes in a day. These data can be used to calculate the total distance it has covered. This information helps identify injury, sickness and stress. Accelerometers, which are closely related to pedometers, are used to measure acceleration forces when an animal moves. They have been proven to be highly accurate in monitoring animal activities and movements.

Respiratory rate sensors and cardiac frequency sensors that typically consist of a belt around the chest are used to measure thoracic/abdominal movements and heart rate, respectively. Abnormal readings indicate the animal is in distress or pain. A noseband sensor is attached to the nose and it monitors eating and ruminating activities in dairy cows. This can help farmers to identify and manage stressed or sick animals.

(v) Sensors in the animal

Immunosensors are used to analyse the content of saliva and sweat to provide an evaluation of hormones in the animal's fluids. Wireless intraruminal bolus (rumen bolus sensor) sensors are inserted to monitor the temperature and pH inside the animal. They can help detect diseases such as those caused by too many acids and nutrients. Photoplethysmography (PPG) detects changes in blood volume using infrared lights.

(vi) Non-attached sensors

Water flow sensors are a type of sensor that is used to monitor the flow of water. They can be used to monitor the drinking pattern of the whole herd.

The scale is used to measure the weight of each animal while cameras can be used to monitor the behaviour of the animals. Sensors in the automatic milking solution (AMS) provide the farmer with milking information like the amount of milk produced by a cow, the quality of the milk, etc.

Using these sensors and several more in combination with each other can be used to monitor and detect a number of possible variables like multiple points of stress, disease or physical pain.

(vii) Automation of rearing work

As indicated in Fig. 3.2, automation of rearing work is one of the elements of SDF. This element is concerned about the raising of the animals with particular interest in milk production. As indicated in the figure, activities include milking, auto feeding, bedding and manure moving. All these processes can be monitored and controlled by using IoT devices. The automation of rearing work ensures the monitoring of the general milk quality, assuring that milk is of high quality. Through the collection of massive amounts of data around the farm, the automation of rearing work provides information for informed decisions to help with the prevention of disease and long-term management of the farm and the animals (Martins et al. 2019).

Modern milking robots have sensors that can detect altered milk properties (e.g., conductivity, colour), hence they have the potential to diagnose mastitis cases.

Technology has improved quality, health and reproduction control in dairy farms. The major benefits of the use of technology include reduced costs of production, better quality of products, better efficiency in production and management, reduced environmental effects and better health and general well-being of dairy animals.

## 3.3 Innovations

In smart dairy farming, innovation can be applied in milk production (milk yield) and the general dairy farming process (Okbar et al. 2019). The major activity in "production" is milking while "process" has a number of activities like smart monitoring, feeding, reproduction and cow observation. Table 3.1 shows the production and process innovations in SDF.

## 3.4 Innovations in Smart Dairy Farming

### A. Digital twins

A digital twin can be described as a virtual imitation of a real-world entity, which simulates the physical, biological and behavioural states of the real-world entity based on input data. It is used in predicting, optimising and improving decision-making

**Table 3.1** Innovations in SDF

| Innovations in SDF | |
|---|---|
| Product | Process |
| **Milk**<br>• *Milking* (robotics used in milking)<br>• *Processing of milk* (measures the milk quality after milking, technology checks colour, temperature, flow and quantity, and this is all recorded and logged (Hansen et al. 2020; Lessire et al. 2020; Reinemann et al. 2021)<br>• Preservation of milk (supercooling technologies to keep the milk fresh for long)<br>• Distribution of milk (traceability as provided by blockchain technology. Food traceability in dairy farming captures, stores and transmits information about milk production in the food supply chain. It ensures milk is checked for safety and quality control traceable in any direction at any time (Casino et al. 2020). IoT and blockchain enable the monitoring of resources and tracking in the value chain. This enables farmers to optimise processes, trace the origin of the produt and guarantee its quality (Alonso et al. 2020) | **Cow observation**<br>• *Activity observation* (uses pedometer, accelerometer, heart rate monitor, cameras, etc. to observe the activity of a cow. Irregular activities will indicate a problem then the farmer can intervene (Lokhorst 2018; Rensis and Scaramuzzi 2003; Mesgaran et al. 2021). Heat stress causes animals to be less active (Waked 2017))<br>• *Behaviour observation* (most of the technologies used for activity observation can also be used here in addition to other sensors to sense the behaviour of an individual cow. A cow can change behaviour due to illness, pregnancy, discomfort, etc. (Lokhorst 2018; Diosdado et al. 2015))<br>• *Body temperature* (temperature sensors are used to monitor the core temperature of a cow. This information, together with input from other sensors, can be used to determine if a cow is on heat or is ill (Awasthi et al. 2016; Helwatkar et al. 2014)) |
| | **Reproduction**<br>• *Reproductive management* (Sensors can be used to determine if a cow is on heat. Using previous data, a farmer can determine the best bull for mating to improve on the breed (Lokhorst 2018)) |
| | **Smart monitoring**<br>• *Positional location* (geographical positioning systems can be used to determine the whereabouts of a certain cow or the whole herd. This helps with management of the flock (Lokhorst 2018; Akhigbe et al. 2021)) |
| | **Feeding**<br>• *Nutritional management* (the farmer is able to tell how much an individual cow is eating and drinking, how much special feed is required, the frequency of feeds, etc. (Da Rosa Righi et al. 2020) |

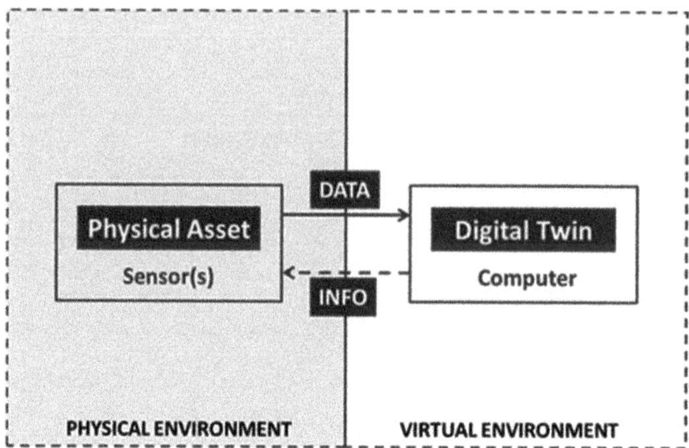


**Fig. 3.5** The digital twin technology concept

(Neethirajan and Kemp 2021). Digital twin technology uses artificial intelligence (AI) to improve efficiency and reduce costs in a number of fields. Recent research has seen digital twin technology being analysed for livestock farming (Verdouw et al. 2021). Digital twins can bring sustainable farming for large-scale productivity. Figure 3.5 illustrates the digital twin concept. A digital twin mirrors the deportment and disposition of a real-world object allowing the real-world object to be controlled and manipulated remotely.

**How digital twins work?**

In the actual (physical) environment, we have the physical asset being monitored. With reference to Fig. 3.5, this asset has sensors attached to it to collect data. The virtual environment is represented by a digital twin in the form of a computing device, e.g. computer, tablet, smartphone. Data collected from the physical asset through sensors are passed onto the digital twin. The digital twin uses algorithms and programs to analyse the data. This analysis is "passed" back to the physical asset as information or decisions, which can then be taken to rectify any problem on the actual asset.

The use of digital twins in dairy farming has the potential to improve precision livestock farming practices and the general well-being of dairy cows. Recognition technology that analyses an animal's facial features (e.g. ear posture) can be used to monitor the emotional and mental states of animals. Using digital twins with other technologies like simulation and augmented reality can help dairy farmers improve their milk production. For example, digital twins can predict animals with disease, negative behaviour of animals and also help farmers construct energy-efficient sheds. With the use of digital twins, farmers can take care of operations remotely using real-time data instead of relying on physical observation and manual tasks on the farm. This lets the farmers respond quickly in case of (expected) deviations and to play out the effects of their responses based on real-life and real-time data.

New research shows the implementation of digital twins in dairy farming. This has helped improve the yield and overall quality of milk and animal health as well as the general management of the farm.

**B. Detection of mastitis**

Mastitis is an udder infection in cows. An affected cow has inflamed mammary glands, and the milk produced from such a cow is not fit for human consumption (Ashraf and Imran 2018). Mastitis results in poor quality of milk, reduced milk production and the slaughter of infected animals. According to Martins et al. (2019), the world dairy industry loses an average of 30 billion euros per annum due to mastitis. It is because of this reason that early and correct diagnosis for mastitis is crucial. Machine learning and multi-spectral imaging have been used in combination to detect mastitis (Hasbahceci and Kadioglu 2018). The technology is based on the fact that infected cows will show certain forms of swelling and temperature in their udders. The technology uses imaging and machine learning to detect these early indicators of the disease before physical manifestations begin to show. This helps in treating the disease before it causes a lot of damage to the infected cows, the other cows and the milk production.

Martins et al. (2019) came up with an innovative way to diagnose mastitis using biosensors. Biosensors are tools used in analytics and they convert the presence of monitored biological compounds into an electric signal. They use living organisms to detect the presence of chemicals (Bhalla et al. 2016). Biosensors have the advantage of stronger signals and fast response times, hence they can detect the presence of the monitored compounds effectively. A number of solutions have been deployed to detect mastitis. Some approaches are based on the detection of immunological effectors as well as changes in the chemical features of milk (Adkins and Middleton 2018; Ashraf and Imran 2018).

## 3.5 Algorithms

In general, smart dairy farming involves the implementation of a wide variety of sensors and devices, which subsequently create a huge amount of data for the farmers. The data may be in the form of text, audio, video and images. These data may be depicted by extremely large sets of either structured or unstructured datasets, which cannot be handled using traditional approaches. Clearly, we can capitalise on data analytics to make sense of the data wherein advanced Artificial Intelligence (AI) and Machine Learning (ML) algorithms can use these data to analyse and predict abnormalities and notify dairy farmers of detected abnormality. To this end, farmers and researchers anchor on the possibilities of big data to develop innovative systems that assist dairy farmers make their cows healthier, happier and more productive (Li 2019). All these innovative systems and approaches are collectively aimed at offering decision support to the farmer and his management decisions. The author in Cockburn (2020) affirms the fact that commercial sensors and systems currently used as animal monitoring systems have already used some of these systems whose underlying algorithms have not been published. Moreover, currently, little is known about the return on investment that these offer or even the effectiveness of their functionality (Cockburn 2021). Clearly, advanced AI and ML have the capacity to speed up this process by imploring decision-making algorithms. Thus ML algorithms are used to automatically "learn" patterns and make assumptions from data. However, there is clearly little information available about the underlying algorithms despite them presenting the essence of performance in the smooth running of smart dairy farms. Machine learning algorithms have been deployed in farm systems that focus on fundamental functions such as feeding, breeding, pasture management, physiology, health and behavioural analysis of dairy cows. In the context of dairy farming and data analytics, these algorithms are classified by Lokhorst et al. (2019) as supervised, semi-supervised, unsupervised and reinforcement learning. Alternative classifications have tended to contain three classes, namely, supervised, unsupervised and reinforcement learning (Kempenaar et al. 2016). We, however, adopt the classifications according to Lokhorst et al. (2019).

This section focuses on common algorithms that are implemented in modern dairy innovation systems.

**A. Supervised Learning Algorithms**: A class of ML algorithms whereby a target (dependent variable) is to be anticipated from a given set of predictors (independent variables). Supervised learning algorithms thus popularly perform classification and regression tasks (Fabris et al. 2017). Common algorithms found in the category of supervised learning and are discussed in this subsection are exemplified by Support vector Machines (SVM), Random Forest Algorithm, Naive Bayes methods and neural Networks.

(i) **Support Vector Machines**: A supervised learning algorithm is used for classification, clustering and regression problems in the dairy industry. The algorithm learns to assign labels to objects using an example (Noble 2006). In essence, it can also be perceived as an algorithm to maximise a specific mathematical function of data. The algorithm creates a hyperplane, which divides the data into two classes. For example in Mammadova and Keskin (2013), the algorithm has been used to check for the presence of mastitis in dairy cows, the behaviour of dairy cows (Diosdado et al. 2015) and for prediction of metabolism of dairy cows (Xu et al. 2019).

(ii) **Random Forest**: Random Forests (RF) is a machine learning algorithm that has been applied to classification and regression problems (Yao et al. 2013). The random forest comprises numerous decision trees with full growth. The tree does not need to be cut, meaning the fuller the tree the more accurate the result (Lin et al. 2017). The algorithm is most suited for datasets containing many predictors (Zaborski et al. 2017). There are two main sources of randomisation in the RF growing procedure: bootstrap sampling of the training data set and the random subsampling of the predictor variables. Moreover, to decrease the chances of model overfitting, the data set is divided into a training and a validation set, both of which are used to estimate a classification error. Clearly, the RF algorithm has been applied in many prediction scenarios ranging from calving events, bull behaviour, dystocia, mastitis, behavioural states in swiss cattle, faciola hepatical infections, detection of eustrus cycle, milk quality (Frizzarin et al. 2021) and classification of bull semen.

(iii) **Naive Bayes:** Decision trees: This type of algorithm is well suited for classification and regression problems. The algorithm leverages tree representation to solve a problem wherein a tree comprises nodes (root, internal and leaf) and edges. The root node has no incoming edges while all the other nodes have exactly one incoming edge. An internal node has outgoing edges, and the other nodes are called leaves. In the learning process, the samples in each interior node are partitioned into subsets in tandem with the value of the attribute. The process, known as recursive partitioning, is repeated in each derived node. The recursion under any of the following three conditions: (i) when a subset of the sample at one node has the same target value, (ii) when splitting no longer improves prediction, (iii) or when splitting cannot be done because of user-defined constraints (Rodríguez et al. 2019).

**B. Unsupervised Learning Algorithms**

Clustering method is generally a ML technique that groups data points. The method deploys algorithms to analyse large quantities of structured and unstructured data. Dave and Gianey (2016) classify the clustering algorithms as follows:

(i) **K-means Clustering**: It is a clustering algorithm that takes K as an initial cluster centre (input parameter). It then divides a set of given objects into K clusters and works out cluster similarity according to the average value of the objects in the cluster. It then assigns each object to the cluster to which it is most similar then calculates the new mean for each cluster (Bangui and Buhnova 2018). This algorithm has been used in a number of scenarios such as grouping the lactation curve (LC) of Holstein cows (Lee et al. 2020), analysing the metabolism of holstein cows (Grelet et al. 2019)

(ii) **Hierarchical based**: Hierarchical clustering is an algorithm that builds a hierarchy of clusters. It starts with all the data points assigned to one cluster then the nearest clusters are merged into one cluster. In the end, there will be only one cluster left (Aggarwal and Reddy 2013)

(iii) **Density based**: The density-based algorithm discovers clusters of different shapes and sizes from large datasets. Each cluster is represented by a set of density-connected objects, which are divided using the region of density, connectivity and boundary (Amini et al. 2014; Guo et al. 2015). The algorithm has high computational complexity hence it is able to improve the communication cost (Bangui et al. 2018).

(iv) **Grid based**: These clustering algorithms are good for dealing with large multi-dimensional data sets. They partition the data into a finite number of cells, forming a grid structure. Clusters are then formed from the cells in the grid structure. Clusters are built from regions that are denser in data points than their surroundings. Currently, literature has no record application of these algorithms to dairy datasets (Aggawal and Reddy 2013).

(v) **Model based**: Model-based clustering algorithms designed for modelling an unknown distribution using a mixture of simpler distributions (Lai et al. 2018). These are premised on the data being generated by probability distributions such as multivariate normal distributions (Yeung et al. 2001).

The advantages of model-based clustering are enumerated by the authors in Devos (2007) as follows:

- The clustering allows for inference due to statistical basis.
- Unlike hierarchical clustering, model-based clustering uses several criteria to assess the optimal number of clusters.
- Incorporates noisy objects, hence no data are lost.
- Visualisation of the cluster shapes is possible using original variables.

These algorithms find applications in identifying dairy topologies and trends (Gonzalez-Mejia et al. 2018).

(vi) **Principal Component Analysis (PCA)**: Reduces a complex data set to fewer dimensions to show the hidden, simplified structures (Shlens 2014). The PCA space consists of orthogonal principal components, i.e. axes or vectors (Tharwat 2016). According to Shlens (2014), the following steps are to be followed:

1. Obtain data
2. Remove the mean
3. Calculate the co-variance matrix
4. Calculate the eigenvectors and eigenvalues of the covariance matrix
5. Choose components to form a feature vector

**C. Semi-supervised learning**

(i) **Transfer learning**: A widely used deep learning technique (Velasco et al. 2020) in small datasets. The usual transfer learning approach is to train a base network and then copy its first selected layers to the first selected layers of a target network. The remaining layers of the target network are then initialised in a random way and trained to achieve the task of classifying cow images by BCS. Two principal techniques known as feature extraction and fine-tuning can be used to apply transfer learning, (Rodriguez et al. 2019). Transfer learning finds applications in predicting pregnancy, determining BCS and evaluation of semen status (Brand et al. 2021).

(ii) **Co-training:** Used to solve the data scarcity problem (i.e. the lack of labelled examples) in supervised learning (Du et al. 2010). It requires two views of data and tends to first learn a separate classifier for each view using labelled examples.

**D. Reinforcement Learning**

Reinforcement learning (RL) is an area of machine learning that uses intelligent agents to act in an environment. In this section, we briefly focus on Markov Decision Process (MDP) and Q-learning.

(i) Markov Decision Process (MDP)

A reinforcement learning technique is used to solve sequential decision-making problems under uncertainty. It is used in decision support systems, for example in herd management systems (Kristensen 2003).

(ii) Q-Learning

Q-Learning is a Reinforcement Learning technique based on learning an action-value function that calculates the expected utility of performing a given action in a particular state and then following a predetermined policy (Manju and Punithavalli, 2011). This algorithm estimates the value of state-activity pairs, and it has three components: environment state, agent action and environment reward.

Q-Learning creates a Q-table, Q(s, a), that indexes a Q-value using state-action pairings. State(s) and reward(r) are updated in Q-Learning at every possible step. The goal of a reinforcement learning problem is defined by the reward function. It assigns a single number, a reward, to each perceived state of the environment, representing the intrinsic desirability of that state. The main goal of a Reinforcement Learning agent is to maximise the total reward it receives over time. The agent's good and bad occurrences are defined by the reward function. In Q-Learning, the agent and the environment interact, and the agent must go through a series of trials to determine the optimum action. An agent chooses the activity that yields the greatest return from its surroundings. Depending on the surroundings, the reward signal may be positive or negative (Manju and Punithavalli 2011) (Table 3.2).

## 3.6 Deployment and Challenges

In the quest to meet the ever-growing demand for increased milk volumes and high-quality dairy products, the Smart Dairy Farming industry has seen new innovations, advancements and subsequently a lot of new startups bringing variety to the dairy ecosystem. The innovations are a result of leveraging sensor technologies, Internet of Things (IoT), cloud computing machine learning and data analytics so as to gain competitive advantage. The leveraging of these technologies and deployment of the ML algorithms results in new systems or technologies that strive to increase productivity, profitability and efficiency of the smart dairy industry operations. In this section, we detail some of the deployment scenarios through Table 3.3.

- **Need for large capital investment**—One of the smart dairy industry's triumphs is the employment of milking robots, which minimise labour demands on dairy

**Table 3.2** Machine learning algorithms deployed in farm systems

| Algorithm | Application |
|---|---|
| K-means, Clustering | Farm classification, marketing, government policy |
| Support Vector Machine (SVM) | Improve water and electricity usage forecasting on pasture-based dairy farms |
| Artificial Neural Networks (ANN) | First test day milk yield |
| Principal Component Analysis (PCA) and Correspondence Analysis (CA) | Farmers can use this classification to make more informed feeding, culling and breeding decisions as they learn more about the health, performance and reproductive qualities of individual cows |
| Convolutional Neural Network (CNN) | Body condition score (BCS) |
| Deep learning algorithm | Detect lame dairy cows |
| Neural networks and random forest | Predicting the rate of breathing, the temperature of the skin and the temperature of the vaginal canal<br>Automated prediction of mastitis |
| Autoregressive integrated moving average models and CA | Detecting diseases such as mastitis |
| Decision tree and random forest<br>Logistic regression | Spread of infectious diseases<br>Herd management |

farms of all sizes and provide a more flexible lifestyle for farm families milking a big number of cows. In the milking process, the milking robot sensors measure variables such as milk temperature, milk yield or the conductivity of milk with algorithms such as random forest being applied to the data leading prediction of cell count, which subsequently facilitates the early detection of infection diseases like mastitis. Regrettably, the deployment of state of the art milking robots sadly requires large capital investment and good management to succeed. The same random forest algorithm can be used in the development of prediction models of dairy farm profitability (Maria and Tauriainen 2014).

- **Datasets and quality**—ML algorithms require huge dairy datasets to make accurate predictions, and these are currently not available. Moreover, the need for quality has always been an issue, and this is made more challenging by the need for real-time big data.
- **Data ownership coupled with privacy and security issues**—Need careful consideration and attention otherwise this may hinder new innovations and development.
- **Integration of Big Data sources**—Integration of numerous heterogeneous data sources is a crucial and challenging issue as this impacts business models.
- **Research and innovation**—The sustainability and profitability of the dairy industry can only be enhanced by adoption of new technologies and this demands requisite skills, research funding and policy shift from authorities especially in the developing world

**Table 3.3** Deployment scenarios

| System/Technology | Place | Description |
|---|---|---|
| Milking robot | Norway, UK, Canada, New Zealand | Cows are ushered onto the robotic milking area wherein an IOAT system decides whether a cow needs to be milked or not. If the cow is to be milked the robotic arm. It then measures the milk from each quarter, its colour, temperature, flow and quantity, and this is all recorded and logged (Hansen et al. 2020; Lessire et al. 2020) |
| Rotary milking system | Germany, USA | Once a cow enters into the robotic milking stall, her number is read from the transponder she wears around her neck. As the milk operation advances, each one of the segments (pieces of pie) representing the cow, changes colour to indicate what stage the milking operation is in for that cow (Reinemann et al. 2021) |
| Dairy Production Analytics (DPA) | Russia: Voshazhnikovo farm | A digital twin of dairy production with the ability to predict collective production for the herd and animal diseases. Key parameters are milking efficiency, diet, health and reproduction (Hooijdonk 2020; Woolley 2021) |
| Dairy Monitor | Netherlands | Also a digital twin is capable of remotely monitoring cows and recognising when a cow is in estrus (heat) and monitoring its health throughout that time. These digital twins also show how the animal behaves through its different phases. The entire procedure provides a full insight into the cow's health and also forecasts when the next cycle will begin dates (Rashid 2018) |

(continued)

**Table 3.3** (continued)

| System/Technology | Place | Description |
|---|---|---|
| Herd Management | India | Numerous gadgets and devices have been developed not only to enhance milk production but to also manage the dairy herd. Examples include drones, rotating swinging brush, collar technology (Singh 2021) |
| 3D Printing: | India, Netherlands | The replacement of machine parts in rural farmers is the main application that will save not only time but also money (Eagle 2016) |
| Energy | Netherlands | A unique dataset on farm energy use was obtained with more than 25,000 observations throughout the years 2015–2018 thanks to an online program that systematically tracked the energy performance of dairy producers (Moerkerken et al. 2021) |
| Grazing | Sweden | Grazing plays a vital role in milk production in both the developed and developing world. The emphasis on production per ha of grazed pasture in New Zealand has led to the development of large herds of intensively grazed spring-calved dairy cattle, often on irrigated pasture to maintain forage production throughout the grazing season, with minimal inputs of silage and concentrate feeds (Wilkinson et al. 2020) |
| Climate smart dairy systems | East Africa | Improved forages and feeding strategies to increase productivity, nutrition and adaptive capacity of smallholder livestock systems (Paul et al. 2017) |

(continued)

- .**Platforms: Developing**—Availing open platforms will help accelerate development of new solutions and innovations and thus subsequently empower the dairy farmer.
- **Connectivity**—In the developing world, internet connectivity is an issue that will need attention to facilitate smart dairy farming. It may clearly be beneficial to

**Table 3.3** (continued)

| System/Technology | Place | Description |
|---|---|---|
| Cow toilet (housing) | Netherlands | Dairy farmers are required to investigate and apply feasible ammonia emission reduction strategies. Dairy production creates a lot of ammonia, especially from 15 to 20 L of urine a cow produces each day. An ingenious system that consists of a feeding station and a urine collection facility. An external stimulation is utilised to stimulate the urination reflex at the end of feed output, and the pee is subsequently collected. The Hanskamp CowToilet collects urine directly and separately from the cow in a unique and intelligent method, causing minimal stress to the anima (Mcculloch 2020) |
| Decision support (dairy brain) | University of Wisconsin-Madison | (1) Nutritional grouping provides a more accurate diet to lactating cows by automatically allocating cows to pens according to their nutritional requirements aggregating and analysing data streams from management, feed, Dairy Herd Improvement (DHI) and milking parlour records; (2) early risk detection of clinical mastitis (CM) that identifies first-lactation cows under risk of developing CM by analysing integrated data from genetic, management and DHI records and (3) predicting CM onset that recognises cows at higher risk of contracting CM (Cabrera and Fadul-Pacheco 2021) |
| Dystocia | Poland | Random forest algorithm is used to detect dystocia (Zaborski et al. 2017) |

(continued)

**Table 3.3** (continued)

| System/Technology | Place | Description |
|---|---|---|
| Phenomics | Scotland, South Africa | Larger dairy farms have switched to automated milking systems, which save time and allow for the automated recording of a huge number of phenotypic (Visser et al. 2020) |
| SmartCow | Europe | SmartCow connects important European cow research infrastructures to encourage their coordinated usage and development, assisting the European cattle industry in meeting the problem of long-term sustainability. SmartCow brings together strong scientific and technical expertise in animal nutrition (in vivo methods for nutrient utilisation and emissions measurements), genetics (genotyped animals, phenotyping capabilities), health and welfare (sensors and automatic recordings of physiological and behavioural parameters) and animal experimentation ethics (Mesgaran et al. 2021) |
| Connected Cow | China | Using real-time monitoring and big data analytics, dairy farmers can assess the health of their cows. Dairy farmers will be able to follow cow estrus in a more precise manner, resulting in increased productivity, efficiency and profitability (Young et al. 2018) |
| Robots for herd Management | Taiwan | In the dairy farm, five working lines using robots were developed to do smart farming as follows: 1. daily milking line, 2. daily cow feeding line, 3. daily cow feces and environmental clean-up line, 4. cycle management of cow calving and young calf feeding line and 5. cycle monitoring of cattle health line for cows and heifers (Wu et al. 2019) |

researchers to consider cost-effective solutions such the works in Nleya et al. (2013), Nleya et al. (2014) and Chiaraviglio et al (2016).

## 3.7 Conclusion

The eradication of hunger is among the greatest global challenges currently facing the world and is a prerequisite for sustainable development. Smart Dairy Farming (SDF) offers opportunities for agriculture to leverage new and upcoming technologies to mitigate this global hunger problem. SDF is mitigating the global hunger problem by deploying innovative solutions made possible through deployment of systems derived from a combination of Artificial intelligence, sensors, Internet of Things, data analytics and connectivity. Thus, the innovative solutions systems do not only increase milk production with regards to the yield but also enhance the process of dairy farming. Some of these notable innovative solutions systems include automatic milking equipment that has the capacity to process and preserve the milk. Moreover, these solutions are also designed to ensure the health and welfare of the very cow producing the milk as they can monitor cow movement, feed and health.

## References

Adkins PRF, Middleton JR (2018) Methods for diagnosing mastitis. Vet Clin North Am 34:479–491. https://doi.org/10.1016/j.cvfa.2018.07.003

Aggarwal CC, Reddy CK (2013) Data clustering, Chapman and Hall/CRC, ISBN: 9781466558229

Akbar MO, Khan MSS, Ali MJ, Hussain A, Qaiser G, Pasha M, Pasha U, Missen MS, Akhtar N (2020) IoT for development of smart dairy farming. J Food Qual. Article ID 4242805, 8 p, 2020. https://doi.org/10.1155/2020/4242805

Akhigbe BI, Munir K, Akinade O, Akanbi L, Oyedele LO (2021) IoT technologies for livestock management: a review of present status, opportunities, and future trends. Big Data Cogn Comput 5:10. https://doi.org/10.3390/bdcc5010010

Alonso RS, Sittón-Candanedo I, García O, Prieto J, Rodríguez-González S (2020) An intelligent edge-IoT platform for monitoring livestock and crops in a dairy farming scenario. Ad Hoc Netw 98, ISSN 1570-8705, https://doi.org/10.1016/j.adhoc.2019.102047. (https://www.sciencedirect.com/science/article/pii/S1570870519306043)

Ashiq A, Vithanage M (2020) Biochar-mediated soils for efficient use of agrochemicals. In: Agrochemicals detection, treatment and remediation, ScienceDirect, https://www.sciencedirect.com/book/9780081030172/agrochemicals-detection-treatment-and-remediation

Ashraf A, Imran M (2018) Diagnosis of bovine mastitis: from laboratory to farm. Trop Anim Health Prod 50:1193–1202. https://doi.org/10.1007/s11250-018-1629-0

Awasthi A, Awasthi A, Riordan D, Walsh J (2016) Non-invasive sensor technology for the development of a dairy cattle health monitoring system. Computers 5:23. https://doi.org/10.3390/computers5040023

Bangui H, Ge M, Buhnova B (2018). Exploring Big Data clustering algorithms for internet of things applications. In: IoTBDS, pp 269–276

Bauerdick J, Treiber M, Hohendinger M, Hijaz O, Schlereth N, Bernhardt H (2019) Connectivity for IoT solutions in integrated dairy farming in Germany. In: Conference: 2019 ASABE annual international meeting at: Boston. https://doi.org/10.13031/aim.201900561

Bhalla N, Jolly P, Formisano N, Estrela P (2016) Introduction to biosensors. Essays Biochem 60(1), 1–8. https://doi.org/10.1042/EBC20150001

Brand W, Wells AT, Smith SL, Denholm SJ, Wall E, Coffey MP (2021) Predicting pregnancy status from mid-infrared spectroscopy in dairy cow milk using deep learning. J Dairy Sci 104(4):4980–4990

Brown TX, Pietrosemoli E, Zennaro M, Bagula A, Mauwa H, Nleya SM (2014, November) A survey of TV white space measurements. In: International conference on e-infrastructure and e-services for developing countries. Springer, Cham, pp 164–172

Cabrera VE, Fadul-Pacheco L (2021) Future of dairy farming from the Dairy Brain perspective: Data integration, analytics, and applications. Int Dairy J, 105069

Casino F, Kanakaris V, Dasaklis TK, Moschuris S, Stachtiaris S, Pagoni M, Rachaniotis NP (2020) Blockchain-based food supply chain traceability: a case study in the dairy sector. Int J Prod Res. https://doi.org/10.1080/00207543.2020.17892

Chiaraviglio L, Blefari-Melazzi N, Liu W, Gutierrez JA, Van De Beek J, Birke R, …, Wu J (2016, November) 5G in rural and low-income areas: Are we ready? In: 2016 ITU Kaleidoscope: ICTs for a Sustainable World (ITU WT), pp 1–8. IEEE

Choi D, Jung S, Lee D, Kim H, Tsang YF, Kwon EE (2021) A new upgrading platform for livestock lignocellulosic waste into syngas using $CO_2$-assisted thermo-chemical process. Energy Convers Manag 236:114084, ISSN 0196-8904. https://doi.org/10.1016/j.enconman.2021.114084

Cockburn M (2021) Can algorithms help us manage dairy cows? 41. GIL-Jahrestagung, Informations-und Kommunikationstechnologie in kritischen Zeiten

Cockburn M (2020) Application and prospective discussion of machine learning for the management of dairy farms. Animals 10(9):1690

Da Rosa Righi R, Goldschmidt G, Kunst R, Deon C, da Costa CA (2020) Towards combining data prediction and internet of things to manage milk production on dairy cows. Comput Electron Agric 169:105156, ISSN 0168-1699, https://doi.org/10.1016/j.compag.2019.105156. (https://www.sciencedirect.com/science/article/pii/S0168169919314437)

Diosdado JAV, Barker ZE, Hodges HR, Amory JR, Croft DP, Bell NJ, Codling EA (2015) Classification of behaviour in housed dairy cows using an accelerometer-based activity monitoring system. Animal Biotelemetry 3(1):1–14

Du J, Ling CX, Zhou ZH (2010) When does containing work in real data? IEEE Trans Knowl Data Eng 23(5):788–799

Eagle J (2016) [https://www.dairyreporter.com/Article/2016/04/21/Dairy-farmer-to-print-3D-cheese) Accessed 21 Apr 2021

Fabris F, De Magalhães JP, Freitas AA (2017) A review of supervised machine learning applied to ageing research. Biogerontology 18(2):171–188

Fournel S, Rousseau AN, Laberge B (2017) Rethinking environment control strategy of confined animal housing systems through precision livestock farming. Biosyst Eng 155:96–123. ISSN:1537-5110, https://doi.org/10.1016/j.biosystemseng.2016.12.005. (https://www.sciencedirect.com/science/article/pii/S153751101630318X)

Frizzarin M, Gormley IC, Berry DP, Murphy TB, Casa A, Lynch A, McParland S (2021) Predicting cow milk quality traits from routinely available milk spectra using statistical machine learning methods. J Dairy Sci

Gonzalez-Mejia A, Styles D, Wilson P, Gibbons J (2018) Metrics and methods for characterizing dairy farm intensification using farm survey data. PloS One 13(5):e0195286

Grelet C, Vanlierde A, Hostens M, Foldager L, Salavati M, Ingvartsen KL, ... Gplus E Consortium (2019) Potential of milk mid-IR spectra to predict metabolic status of cows through blood components and an innovative clustering approach. Animal 13(3):649–658

Guarino M, Norton T, Berckmans D, Vranken E, Berckmans D (2017) A blueprint for developing and applying precision livestock farming tools: a key output of the EU-PLF project. Anim Front 7(1):12–17. https://doi.org/10.2527/af.2017.0103

Hansen BG, Bugge CT, Skibrek PK (2020) Automatic milking systems and farmer wellbeing–exploring the effects of automation and digitalization in dairy farming. J Rural Stud 80:469–480

Hasbahceci M, Kadioglu H (2018) Use of Imaging for the diagnosis of Idiopathic granulomatous mastitis: a clinician's perspective. J Coll Physicians Surg Pak 28(11):862–867. https://www.cpsp.edu.pk/jcpsp.pk/archive/2018/Nov2018/12.pdf. Accessed 18 Apr 2021

Helwatkar A, Riordan D, Walsh J (2014) Sensor technology for animal health monitoring. In: Proceedings of the 8th international conference on sensing technology, Liverpool, UK

Hooijdonk RV (2020) IoT technology is transforming the agricultural sector as we know it (https://www.iotforall.com/big-data-in-agriculture)

Kristensen AR (2003) A general software system for Markov decision processes in herd management applications. Comput Electron Agric 38(3):199–215

Kulatunga C, Shalloo L, Donnelly W, Robson E, Ivanov S (2017) Opportunistic wireless networking for smart dairy farming. IT Professional 19(2):16–23. https://doi.org/10.1109/MITP.2017.28

Lessire F, Moula N, Hornick JL, Dufrasne I (2020) Systematic review and meta-analysis: identification of factors influencing milking frequency of cows in automatic milking systems combined with grazing. Animals 10(5):913

Lee M, Lee S, Park J, Seo S (2020) Clustering and characterization of the lactation curves of dairy cows using K-medoids clustering algorithm. Animals 10(8):1348

Li L (2019) Cargill: revolutionizing dairy farming with magic big data (https://digital.hbs.edu/platform-digit/submission/cargill-revolutionizing-dairy-farming-with-magic-big-data/). Accessed 18 Apr 2021

Lin W, Wu Z, Lin L, Wen A, Li J (2017) An ensemble random forest algorithm for insurance big data analysis. IEEE Access 5:16568–16575

Lokhorst C (2018) An introduction to smart dairy farming. Van Hall Larenstein University of Applied Sciences, 108 p

Lokhorst C, de Mol RM, Kamphuis C (2019) Invited review: Big data in precision dairy farming. Animal 13:1519–1528. https://doi.org/10.1017/S1751731118003439

Long X, Sun C, Tan M (2020) Design and Implementation of Intelligent Ear Tag for Dairy Cows in Farms. In: Proceedings of the 2020 9th international conference on software and computer applications (ICSCA 2020). Association for Computing Machinery, New York, NY, USA, pp 297–301. https://doi.org/10.1145/3384544.3384574

Lovarelli D, Bacenetti J, Guarino M (2020) A review on dairy cattle farming: Is precision livestock farming the compromise for an environmental, economic and social sustainable production? J Clean Prod 262:121409, ISSN 0959-6526, https://doi.org/10.1016/j.jclepro.2020.121409. (https://www.sciencedirect.com/science/article/pii/S0959652620314566)

Mammadova N, Keskin I (2013) Application of the support vector machine to predict subclinical mastitis in dairy cattle. Sci World J

Manju S, Punithavalli M (2011) An analysis of Q-learning algorithms with strategies of reward function. Int J Comput Sci Eng 3(2):814–820

Martins SAM, Martins C, Cardoso FA, Germano J, Rodrigues M, Duarte C, Bexiga R, Cardoso S, Freitas PP (2019) Biosensors for on-farm diagnosis of mastitis. Front Bioeng Biotechnol 7:186. https://www.frontiersin.org/article/. https://doi.org/10.3389/fbioe.2019.00186, ISSN 2296–4185

Mcculloch C (2020) Dutch cow toilet wins gold medal at EuroTier. Food Farming Technol Mag (https://www.foodandfarmingtechnology.com/news/environment/dutch-cow-toilet-wins-gold-medal-at-eurotier.html). Accessed 11 Apr 2021

Moerkerken A, Duijndam S, Blasch J, van Beukering P, Smit A (2021) Determinants of energy efficiency in the Dutch dairy sector: dilemmas for sustainability. J Clean Prod 293:126095

Neethirajan S, Kemp B (2021) Digital twins in livestock farming. Animals 11:1008. https://doi.org/10.3390/ani11041008

Nleya SM (2016) Design and optimisation of a low cost Cognitive Mesh Network

Nleya SM, Bagula A, Zennaro M, Pietrosemoli E (2013, October) A TV white space broadband market model for rural entrepreneurs. In: Global information infrastructure symposium-GIIS 2013. IEEE, pp 1–6

Nleya SM, Bagula A, Zennaro M, Pietrosemoli E (2014) Optimisation of a TV white space broadband market model for rural entrepreneurs. J ICT Stand 2(2):109–128

Noble WS (2006) What is a support vector machine? Nat Biotechnol 24(12):1565–1567

OECD and Food and Agriculture Organization of the United Nations (2020) OECD-FAO Agricultural Outlook 2020–2029. https://doi.org/10.1787/1112c23b-en

Paul BK, Notenbaert AMO, Mutimura M (2017) Climate-smart dairy systems in East Africa: R4D in support of IFAD loan programs

Quan X, Shan J, Xing Y, Peng C, Wang H, Ju Y, Zhao W, Fan J (2021) New horizons in the application of a neglected biomass pyrolysis byproduct: a marked simultaneous decrease in ammonia and carbon dioxide emissions. J Clean Prod, 297:126626. ISSN 0959-6526,https://doi.org/10.1016/j.jclepro.2021.126626

Rashid M (2018) Why modern farming need the digital twins. Challenge Magazine (https://www.challenge.org/knowledgeitems/why-modern-farming-need-the-digital-twins/). Accessed 18 Apr 2021

Reinemann DJ, van den Borne BHP, Hogeveen H, Wiedemann M, Paulrud CO (2021) Effects of flow-controlled vacuum on milking performance and teat condition in a rotary milking parlor. J Dairy Sci

Rensis FD, Scaramuzzi RJ (2003) Heat stress and seasonal effects on reproduction in the dairy cow-a review. Eriogenology 60(6), 1139–1151

Rodríguez E, Waissman J, Mahadevan P, Villa C, Flores BL, Villa R (2019) Genome-wide classification of dairy cows using decision trees and artificial neural network algorithms

Shen Z, Zhang Y, McMillan O, O'Connor D, Hou D (2020) Chapter 6—The use of biochar for sustainable treatment of contaminated soils. In: Sustainable remediation of contaminated soil and groundwater. Butterworth-Heinemann, pp 119–167, ISBN 9780128179826, https://doi.org/10.1016/B978-0-12-817982-6.00006-9

Shlens J (2014) A tutorial on principal component analysis. arXiv preprint arXiv:1404.1100

Singh R (2021) Role of IoT in transforming Indian dairy industry to smart dairy farming (https://www.pashudhanpraharee.com/role-of-iot-in-transforming-indian-dairy-industry-to-smart-dairy-farming/)

Taneja M, Jalodia N, Byabazaire J, Davy A, Olariu C, (2019) SmartHerd management: a microservices-based fog computing–assisted IoT platform towards data-driven smart dairy farming. J Softw: Pract Exp. https://doi.org/10.1002/spe.2704, IEEE. IEEE Standards Association—IoT ecosystem study. http://standards.ieee.org/innovate/iot/study.html

Tharwat A (2016) Principal component analysis-a tutorial. Int J Appl Pattern Recognit 3(3):197–240

Tullo E, Finzi A, Guarino M (2019) Review: environmental impact of livestock farming and precision livestock farming as a mitigation strategy. Sci Total Environ 650, Part 2:2751–2760, ISSN 0048-9697,https://doi.org/10.1016/j.scitotenv.2018.10.018

Vasisht D, Kapetanovic Z, Won J, et al (2017) FarmBeats: an IoT platform for data-driven agriculture. In: Proceedings of the 14th USENIX symposium on networked systems design and implementation (NSDI), Boston, MA

Velasco JS, Arago NM, Mamba RM, Padilla MVC, Ramos JPM, Virrey GC (2020) Cattle Sperm classification using transfer learning models. Int J 8(8)

Verdouw C, Tekinerdogan B, Beulens A, Wolfert S (2021) Digital twins in smart farming. Agric Syst 189:103046

Waked AM (2017) Kuwait climate and heat stress in dairy cattle. University of Kuwait, Kuwait City, Kuwait

Wilkinson JM, Lee MR, Rivero MJ, Chamberlain AT (2020) Some challenges and opportunities for grazing dairy cows on temperate pastures. Grass Forage Sci 75(1):1–17

Wolfert S, Ge L, Verdouw C, Bogaardt MJ (2017) Big data in smart farming—a review, agricultural systems. 153:69–80. ISSN-0308-521X,https://doi.org/10.1016/j.agsy.2017.01.023

Woolley RD (2021) https://www.iotjournaal.nl/wp-content/uploads/2021/04/BR_How-to-Succeed-with-IoT.pdf

Wu MC, Chao CW, Shiau JW, Chang HL (2019) Robots for herd management of dairy cows in tropical Taiwan (https://ap.fftc.org.tw/article/1617). Accessed 20 Apr 2021

Xu W, van Knegsel AT, Vervoort JJ, Bruckmaier RM, van Hoeij RJ, Kemp B, Saccenti E (2019) Prediction of metabolic status of dairy cows in early lactation with on-farm cow data and machine learning algorithms. J Dairy Sci 102(11):10186–10201

Yang L, Ge X (2016) Chapter three—biogas and syngas upgrading, advances in bioenergy, Elsevier, Volume 1, Pages 125–188, ISSN 2468-0125, ISBN 9780128095225, https://doi.org/10.1016/bs.aibe.2016.09.003

Yao C, Spurlock DM, Armentano LE, Page CD Jr, VandeHaar MJ, Bickhart DM, Weigel KA (2013) Random Forests approach for identifying additive and epistatic single nucleotide polymorphisms associated with residual feed intake in dairy cattle. J Dairy Sci 96(10):6716–6729

Yeung KY, Fraley C, Murua A, Raftery AE, Ruzzo WL (2001) Model-based clustering and data transformations for gene expression data. Bioinformatics 17(10):977–987

Yong N, Ge H, Tse W and Pang K (2018) (https://www.gsma.com/iot/wp-content/uploads/2018/03/GSMA-IoT-Case-Study-Greater-China-EN-March-2018.pdf). Accessed 15 Apr 2021

Zaborski D, Proskura WS, Grzesiak W, Szatkowska I, Jędrzejczak-Silicka M (2017) Use of random forest for dystocia detection in dairy cattle. Appl Agric For Res, 147

Zin TT, Pwint MZ, Seint PT, Thant S, Misawa S, Sumi K, Yoshida K (2020) Automatic cow location tracking system using ear tag visual analysis. Sensors 20:3564. https://doi.org/10.3390/s20123564

# Chapter 4
# Precision Farming in Modern Agriculture

**E. Fantin Irudaya Raj, M. Appadurai, and K. Athiappan**

**Abstract** The science, art and practice of growing plants are called agriculture. The history of agriculture began thousands of years ago. It began independently in various parts of the world depending on climatic conditions and terrains. Agriculture enabled the human population to expand several times beyond what could be supported by hunting and gathering. From the twentieth century, precision farming in contemporary agriculture increased productivity. Precision farming is a technology-enabled approach to farming that measures, observes and analyses individual fields and crops' needs. This type of farming's main aim compared with traditional farming is where inputs are utilized in precise amounts to increase crop yields and profitability. The present work detailed about advancement and role of the Artificial Intelligence (AI) and Internet of Things (IoT) used in precision farming. The AI and IoT play a vital role in our modern day-to-day applications. The advantages and advancements of AI and IoT are applied in modern agriculture. This makes modern agriculture more precise and profitable. Some of the contemporary applications using AI and IoT in smart precision farming are discussed. In addition to that, important tools and techniques deployed in precision farming are also explained. The crucial benefits and the real-time devices used in the precision farming are also discussed in detail.

**Keywords** Modern agriculture · Internet of Things (IoT) · Artificial Intelligence (AI) · Precision farming

---

E. F. I. Raj (✉)
Department of Electrical and Electronics Engineering, Dr. Sivanthi Aditanar College of Engineering, Thiruchendur, Tamil Nadu, India

M. Appadurai
Department of Mechanical Engineering, Dr. Sivanthi Aditanar College of Engineering, Thiruchendur, Tamil Nadu, India

K. Athiappan
Department of Civil Engineering, Jyothi Engineering College, Thrissur, Kerala, India

 
A. Choudhury et al. (eds.), *Smart Agriculture Automation using Advanced Technologies*, Transactions on Computer Systems and Networks,
https://doi.org/10.1007/978-981-16-6124-2_4

## 4.1 Introduction

The complex interaction of crops, soil, water and pesticides results in the agriculture production system. For the sustainability of such a complex system, prudent management of all inputs is essential. To achieve optimal growth, farmers must understand the cultivation of crops in a specific field, taking into account the resistance of the seeds to local disease and weather conditions and taking into account the environmental effects on the seeds planting. For example, it's better to use a seed that needs less fertilizer when planting near a river, and it helps to minimize the contamination due to fertilizer.

After the seeds have been planted, fertilizing and preserving the crops is a time-consuming process heavily influenced by the weather. Farmers will decide not to apply fertilizer if they predict heavy rain the next day because it will be washed away. Predicting and knowing whether it will rain or not can also help determine when the fields are to be irrigated. Throughout the world, seventy percentage of the freshwater used in agriculture; will have a significant influence on the world's fresh water supply and can be managed in a better way how it is used.

Weather affects not only the growth of crops but also logistics relating to harvesting and transport. The soil must be sufficiently dry to support the weight of the harvesting equipment. When the weather is humid and the soil is muddy, the machinery may destroy the crop. With a clearer picture of the weather and which fields will be affected in the coming days, better decisions can be taken about which field workers should be deployed ahead of time.

Harvesting and transporting crops to supply centres is critical as soon as the crops have been harvested. Food waste is normal during delivery, so keeping food at the correct temperature and not storing it for longer than required is critical. The weather even can have a considerable impact; if the roads are muddy, heavy rain can easily trap trucks. Business owners will make more informed decisions about the fastest routes for transporting their food if they know where it will rain and which routes will be impacted.

Thus, agriculture was traditionally practised by following a predetermined schedule for performing a specific task, such as planting or harvesting. Predictive analytics, on the other hand, will help you make better decisions by collecting real-time data on soil, crop maturity, air quality and weather, as well as equipment and labour costs and availability. Precision Agriculture (PA) is the name given to this form of modernized agriculture. It's a farm management technique that employs Information Technology (IT) and Artificial Intelligence (AI) to ensure that soil and crops obtain the nutrients they required for finest production and quality. Its key goals are to ensure profitability, long-term viability and environmental stewardship.

PA relies on special software packages, specialized equipment and information technology services. Obtaining continuous information about the crop, soil, ambient air conditions and other critical data such as hyper-local weather forecasts, labour costs and equipment availability are all part of the strategy. Predictive analytics software uses the data to provide farmers with advice and recommendations on crop rotation, soil management, harvesting times and optimal planting times.

Sensors in the field calculate the moisture content, temperature and ambient air conditions. Individual real-time images of the plants grown are provided by satellites and robotic drones to cultivators. The data from the digital images received is retrieved and can be analysed and combined with the data received from the signals to provide direction for instantaneous and imminent decisions, such as when or where to cultivate a certain crop, and which fields to water.

Imaging input and sensor data are combined with other data at agricultural control centres to enable cultivators to identify fields in need of treatment and decide the best pesticides, water and fertilizer to use. It will allow the farmer to avoid agriculture run-off and water wastage while also lowering costs and reducing the environmental impact of the farm. PA has traditionally been limited to larger operations with the financial capital to invest in the IT infrastructure and other technical tools needed to fully comprehend and profit from its benefits. Because of the advancement of cloud computing, drones, smart sensors and mobile apps, PA is now possible for small family farms and farming cooperatives.

## 4.2 Related Works

PA is a new term of farm management that has a lot of different definitions. To explain the principles of PA in general, two important definitions have been chosen. In 1997, the United States House of Representatives presented the first definition. It states that, 'An integrated information and production-based farming system that is designed to improve long-term, site-specific and whole farm production quality, productivity, and profitability while mitigating unintended impacts on wildlife and the environment'. The second definition focuses on site-specific crop management systems. It defines PA as 'a form of precision agriculture in which resource application and agronomic practises are enhanced to better match soil and crop requirements as they differ in the field'. This term includes the concept of PA as a dynamic management technique.

The various aspects of PA and the implementation of PA in the twenty-first century are discussed in detail (Pierce and Nowak 1999; Stafford 2000). The principles to manage spatial and temporal availability associated with all aspects of agricultural production to improve crop performance and environmental quality are addressed. The recent technology advancements like the Internet of Things (IoT), AI and Satellite-based global positioning system make precision farming easier in recent years. In (Zhang et al. 2002; Srinivasan 2006; Baggio 2005; Adamchuk et al. 2004; Duhan et al. 2017; Robert 2005; Liaghat and Balasundram 2010; Cox 2002), the author details the worldwide development and the current status of PA technologies. The topics variability management, the impact of PA, management zone and natural resource availability is explained in detail. Innovation in the sensors, remote sensing and controls are detailed. The adoption trend of PA technologies in recent years was also demonstrated.

The IoT is a term that refers to the billions of physical devices connected to the internet on a global scale and collecting and sharing data. Incorporating the IoT in modern agriculture makes it more reliable and profitable. In (García et al. 2020; Keswani et al. 2020; Sanjeevi et al. 2020), the authors detailed the various sensors employed in precision farming. They discussed the different protocols that need to be followed when using IoT devices. An IoT-enabled automatic irrigation system that is driven by big data is introduced. It will modernize real-time modern agriculture. In addition to that, the authors brief about using wireless networks incorporated with IoT for PA and farming. Appropriate utilization of water irrigation management can be achieved by applying WSN technology using IoT. The efficient communication of various wireless sensors is processed using IoT in PAF to improve farmers' productivity. In (Kashyap et al. 2021; Ponnusamy and Natarajan 2021; Sumarudin et al. 2021), the author detailed various models used recently in the automatic irrigation system in modern agriculture. Deep learning and machine learning are recent development in AI world. Many new algorithms proposed to train a model more precisely to train the neural network and make it a more efficient one. The author discussed incorporating the deep learning models and machine learning models like Support Vector Machine (SVM) with IoT and makes the more efficient model of automatic smart irrigation system. They also detailed the usage of augmented reality and unmanned aerial vehicles in modern PA.

AI is a wide range of computer technology, which focuses on the production of smart machines capable of performing normal human-led tasks. While AI is a multidisciplinary science that employs a variety of techniques, advances in machine learning and deep learning are causing a paradigm shift in nearly every sector of the technology industry. By means of applying AI into modern agriculture makes it more profitable and sustainable. The authors detailed the incorporation of IoT and AI in precision farming (Ampatzidis et al. 2020; Wei et al. 2020). They discussed in detail the automation in crop monitoring using remote sensing images and the image of the individual plant. Employing this, the cultivator can monitor the individual plant in the field from remote places even. In addition to that, the remote sensing and AI gives the information about the weather forecast and provide the better ideas of the soil condition to the farmers to make proper decision to cultivate the right plants in the right time. Disease detection and crop yield analysis are also crucial points to be considered in precision farming (Lee et al. 2020; Talaviya et al. 2020; Costa et al. 2020). The authors detailed the yield mapping, machine learning-based satellite-driven vegetation index for the PA and leaf stomata properties using AI and machine vision.

Block chain technology is also one of the crucial technologies recently evolved. Integrating the IoT and Block chain and using it in PA makes it more profitable and sustainable. Land administration is one of the critical factors to be considered in PA. The following is a model for using block chain technology to enter data into the land information system: data from users is obtained and added to the block chain, they are replicated and the transaction is validated, a block with a transaction is created and it is added to the block chain of all system participants (Beznosov et al. 2021). The authors discussed the block chain and edge computing technology to enable the organic agricultural supply chain (Liu et al. 2021). The applications of information and communication technologies and block chain technologies for the development of PA are discussed in detail (Hu et al. 2021).

Several more similar works in the literature explained the conversion of our traditional agriculture into modern PA using these modern technologies. In the present work, the role of AI and IoT in Precision farming, the Role of Nanotechnology in Precision farming, Real-time modern equipment used in Precision farming and the benefits and shortfalls of Precision farming are discussed in detail in the following chapters.

## 4.3 Role of IoT in Precision Agriculture

The agricultural sector must grow to meet its demands, regardless of environmental challenges, such as unfavourable weather and climate change, in view of the increasing population, now expected to reach 9.6 billion by 2050. Agricultural industries must use new technologies to take a much-needed lead to meet the demands of this growing population. Through IoT-enabled agricultural applications such as smart farming and precision farming, the industry will be able to improve operational efficiency, reduce costs, eliminate waste and improve the quality of their yield. In precision farming, the IoT is mainly used in Smart agriculture monitoring system and automated irrigation system.

The IoT-based intelligent monitoring system and automated irrigation system is shown in the following Fig. 4.1. In this system, various sensors have been deployed in the various part of the agriculture field. The important sensors used in the system are rain detection sensor, temperature sensor, water tank level indicator, soil moisture sensor and humidity sensor. The actual placing of these sensors in the agriculture field is as shown in Fig. 4.2. These sensors sense the actual physical value in the ground field and convert in to a digital signal and transmit to the signal processing unit. This unit is connected with the Arduino processor. The processor is equipped with the LCD display for the instantaneous monitoring of the actual value in the field to the observer. The Arduino also sends the suitable command signal to the motor control unit, which directs the motor in the agriculture field. Thus, based upon the information received from the sensors the system can act as an automated irrigation system. The information received can also be transferred through Wi-Fi module through internet connectivity to cloud-based system. From this the farmer or observer can monitor the actual data and take corrective actions from the remote place accordingly. Figure 4.3 shows the web application used in the monitoring system; Using this the farmer can observe the field data from the remote place. In this way, it can act as an effective monitoring system. Thus, the IoT-based automated irrigation system and monitoring system makes the conventional farming into smart and precision farming.

The IoT-enabled smart agriculture systems make farming practises, such as livestock monitoring, vehicle tracking and inventory tracking more precise and controlled. The aim of precision farming is to analyse and react accordingly to data

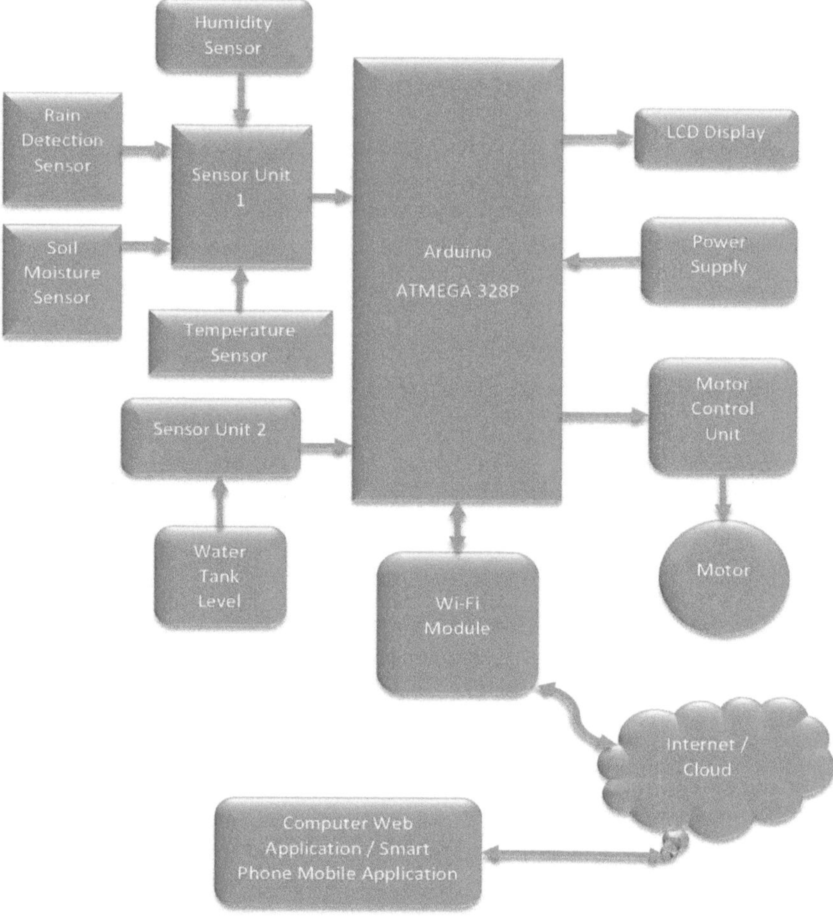


**Fig. 4.1** IoT-based intelligent monitoring and automated irrigation system

generated via sensors. Precision Farming enables farmers to collect data via sensors and analyse it in order to make intelligent and timely decisions. Numerous precision farming techniques exist, including irrigation management, livestock management and vehicle tracking, all of which contribute significantly to increasing efficiency and effectiveness. Precision farming enables you to analyse soil conditions and other relevant parameters in order to maximize operational efficiency. Not only that, but you can also monitor the connected devices' real-time operating conditions to determine their water and nutrient levels.

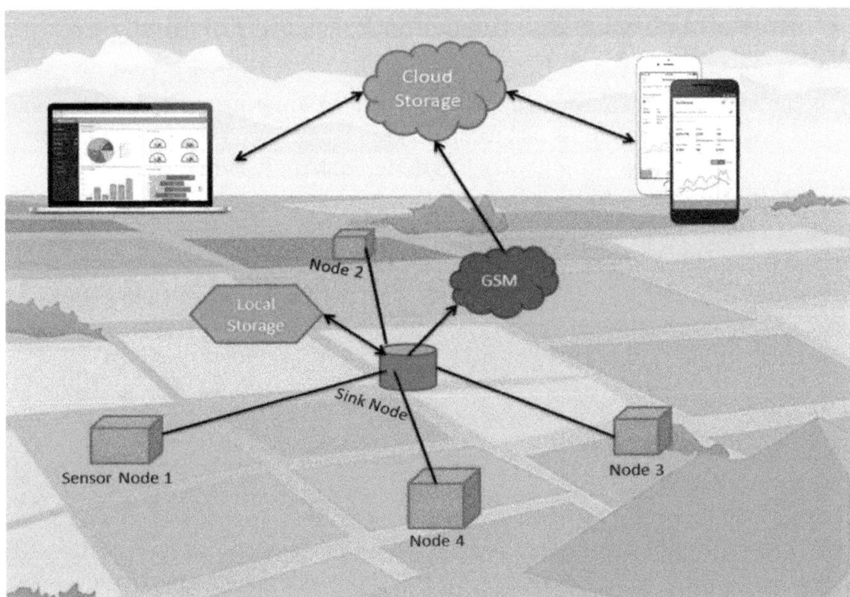


**Fig. 4.2** Actual placing of sensors in the agricultural field

**Fig. 4.3** Web-based application used for monitoring the actual field data by the observer

## 4.4 Role of Artificial Intelligence in Precision Farming

AI is among the most prominent research area of Information technology and computer science research. Its fast technical development and wide range of applications can be extended to traditional farming, making it more precise and cost-effective. The significant issues in traditional agriculture are an inadequate application of chemicals, pest and disease infestation, improper irrigation and drainage, yield prediction, weed control, etc. The application of AI in traditional farming overcome these shortfalls and supports precision farming. The significant areas where AI applied in precision farming are discussed further.

### *4.4.1 Pest Management*

Livestock and crops can be harmed by diseases, insects and weeds, which can be costly and irreversible. Pesticides and biological pest control are two options for dealing with these issues. Researchers have spent decades attempting to reduce this threat by designing computerized systems that can detect active pests and recommend control measures. Since agricultural management information is incomplete, ambiguous and imprecise, the rule-based expert system can cause confusion. The object-oriented approach to establish a rule base for TEAPEST, an expert system for tea pest control is discussed by Ghosh and Samanta (2003). A phase-by-phase recognition and consultation procedure was also used in this case.

To address these shortfalls, many Fuzzy logic-based expert systems have been suggested, including Siraj and Arbaiy (2006), Saini et al. (2002), Peixoto et al. (2015), Roussel et al. (2000), Jesus et al. (2008), Shi et al. (2007) and Hayo et al. (1998). Later, Samanta et al. revamped the device with a neural network with multi-layered back propagation methodology (Samanta and Ghosh 2012) and Banerjee et al. redeveloped it with a radial basis function model to reach better classification rates (Banerjee et al. 2017). The effectiveness of the deep learning-based classifiers in terms of accuracy, sensitivity and specificity are explained in detail (Raj and Balaji 2021).

### *4.4.2 Crop Management*

It deals with the aspects of agricultural productivity such as increasing growth, development and yield, as well as prevention and management techniques for crop improvement. The combination, timing and sequence of practises used are determined by the biological characteristics of the crops (spring or winter crops), the harvested form (green feed, grains and so on), the sowing method (wide-row, row or nest), the soil and the plant age, weather and climatic conditions.

In general, Crop and harvest management systems, offer an interaction with all aspects of farming for full crop management. Mckinion et al. suggested using AI techniques in crop management in their paper 'Expert System for Agriculture' (Bannerjee et al. 2018) in 1985. In his doctoral thesis (Boulanger 1983), Boulanger suggested a new corn crop safety system. In Italy, researchers developed a multi-layered feed-forward network-based system to protect citrus crops from frost damage (Robinson and Mort 1997). To train and evaluate the network, the input and output parameters were coded in binary form. The authors experimented with various input configurations to find the most accurate model. With two output groups and six inputs, the best model found so far had a 94% accuracy. To improve the image transformation, Li, S. K. et al. use a pixel labelling algorithm followed by Laplace transformation for wheat crop (Li et al. 2002). The optimal network obtained consisted of five hidden layers that had been trained for 400,000 iterations and averaged 91.1% of accuracy. C. Prakash et al. developed a fuzzy logic-based soybean crop management framework that provided advice on crop selection, fertilizer application and pest-related issues (Prakash et al. 2013).

### *4.4.3 Soil and Irrigation Management*

Irrigation and soil management may have a significant effect on soil microorganisms and potential seed decay agents, as the concentration of these agents is highly dependent on soil moisture levels. Poor irrigation and soil management leads to quality degradation and crop loss. This part highlights some soil and irrigation management research that has been enabled by AI techniques. To assess the efficiency of micro-irrigation systems, Brats et al. (1993) developed a rule-based expert framework.

Sicat et al. (2005) used farmers' knowledge to design a modern fuzzy-based framework that recommended crops based on the formation of suitable land aptness maps. Tremblay et al. (2010) and Si et al. (2007) are two other fuzzy logic-based systems. On the basis of meteorological and soil water content data, for the purposes of estimation of stem water potential, A Sugeno method-based adaptive fuzzy inference system was used by Valdes-Vela et al. (2015). For estimating soil moisture levels in paddy, Arif et al. (2013) developed a neural network-based system. Broner and Comstock (1997) are two other common artificial neural network-based soil and irrigation systems.

### *4.4.4 Agriculture Product Monitoring and Storage Control*

In addition to pests and disease management, agriculture requires constant monitoring, grading, storage and drying of harvested crops. Various AI-based food monitoring and quality control systems are discussed in this section. Gottschalk et al.

(2003), Escobar et al. (2004) and Kavdir et al. (2004) are only a few of the fuzzy logic-based systems that have been developed. Yang (1993), Taki et al. (2016), Capizzi et al. (2016) and Nakano (1997) are examples of artificial neural network-based approaches that need to be addressed.

### 4.4.5 *Disease Management*

Another major concern for farmers is crop diseases. To diagnose an ailing plant and take the appropriate recovery measures, you'll need a lot of knowledge and experience. Computer-assisted systems are used all over the world to identify illnesses and recommend treatment options. Rule-based systems have been developed at an early stage, it includes Balleda et al. (2014), Sarma et al. (2010), Byod and Sun (1994) and Tilva et al. (2013) suggested a fuzzy-based method for predicting diseases based on the duration of leaf wetness.

Some hybrid systems have also been proposed. To identify phalaenopsis seedling diseases, Huang suggested a model that combines image processing and artificial neural networks (Huang 2007). Sannakki et al. (2011), proposed a system along with an adaptive fuzzy inference-based approach to measure the leaf infection percentage, employed digital image processing technique. Al Bashish et al. (2011) and Al-Hiary et al. (2011) developed a method based on the k-means segmentation algorithm. Khan et al. developed Dr Wheat, a web-based expert system for diagnosing wheat diseases (Khan et al. 2008).

### 4.4.6 *Yield Prediction*

Decision-makers at the national and regional level need to be given sufficient information on the expected yield of crops to enable fast decisions. It is extremely useful for adapting various marketing strategies and estimation of cost. Prediction models may also be used to analyse relevant factors that directly affect yield there in age of precision farming. To predict yield from soil parameters, Liu et al. (2005) proposed back-propagation learning algorithm assisted artificial neural network system. Pahlavan et al. used energy output to calculate yield for basil plants in the greenhouse in a different way (Pahlavan et al. 2012). Nabavi-Pelesaraei et al. (2016), Soheili-Fard et al. (2015) and Khoshnevisan et al. (2013) are some of the other important research works on yield prediction. Rode and Dahikar suggested an artificial neural network model for forecasting the yields of seven distinct crops based on atmospheric inputs and fertilizer usage (Dahikar and Rode 2014).

### 4.4.7 Weed Management

Weed control is essential in all types of agriculture systems. The weed is controlled by several methods such as mechanical, biological, cultural and chemical methods. For example, in mechanical weed control techniques, the primary tillage is done for burying the perennial seeds in depth. Due to this technique, the weed seeds are prevented from emerging with cultivated crops for a certain period. The use of herbicides in chemical means of weed control gave higher production efficiency, less mechanical tillage practice and efficient all types of weed control.

Herbicide application has a significant impact on human health as well as the ecosystem. Via proper and precise weed control, modern AI methods are being used to reduce herbicide usage. Pasqual (1994) developed a rule-based intelligent arrangement for weed detection and elimination in triticale, oats, wheat and barley crops. Burks et al. (2000) used back-propagation trained neural network and a machine vision to identify five different weed species. Three separate neural network models were compared by Burks et al. (2005). Using the same set of inputs as used earlier in the previously mentioned articles, counter-propagation, back-propagation and radial basis function-based models were used to determine. The proposed network performs the best, with a 97% accuracy. Shi et al. (2007) developed another method that combined neural network and image analysis. Other noteworthy works include those by Barrero et al. (2016), Eddy et al. (2008) and Nebot et al. (2012).

## 4.5 Role of Other Important Tools and Techniques Used in Precision Farming

PA is unique in that it does not focus on a single technology to enhance a single practise. PA is emerging as a fusion of many technologies applied to a variety of crop management practises across the crop production market.

However, not every technology is needed or applicable for every practise on all crops and forces outside the agricultural sector are driving the development and enhancement of many potentially relevant basic technologies. As a result, forming a consensus on the dimensions of PA is challenging. Every aspect of information technology, including computers, sensors, computers and microelectronics, is undergoing a continuous improvement phase. Figure 4.4 shows the various technologies employed in the PA.

Many of the innovations used in PA were developed outside of the agricultural community. Other inventions outside of agriculture, such as electrical power, the internal combustion engine, weather satellites and telephones, have all been applied to the agriculture sector over the last century. PA technologies as shown in Fig. 4.4 incorporates various technologies such as Remote Sensing, Geographic Information Systems (GIS) and Global Positioning Systems (GPS) all have significant non-agricultural applications. Soil and Crop sensors mounted on farm machinery, yield

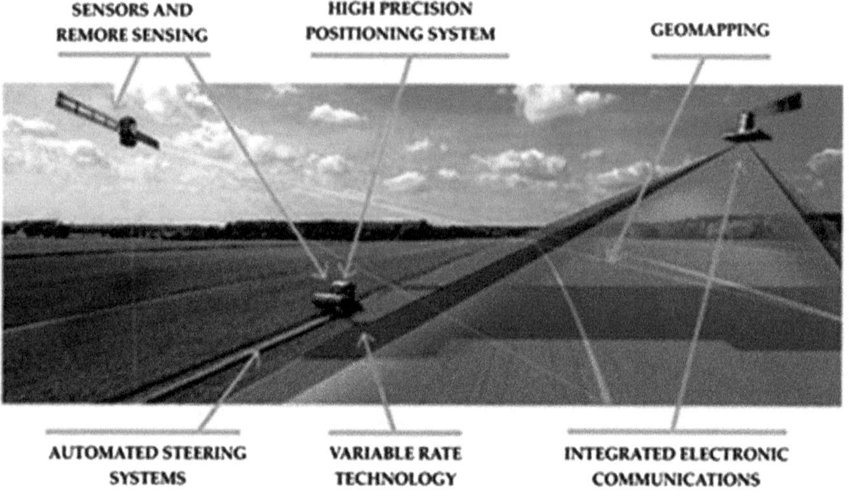


**Fig. 4.4** Various technologies employed in precision agriculture

mapping systems and variable-rate fertilizer applicators are all agricultural technologies developed by private industry. Other economic sectors have aided in the research and development of several of these technologies, which benefits agriculture financially. PA is the process by which these information technologies are integrated with agronomic knowledge.

### *4.5.1 Remote Sensors*

Remote sensing refers to data collected by instruments that detect visible light, infrared light, electromagnetic radiation and near-infrared light without making physical contact with the object of study (Fig. 4.5). Remote sensing can be used to produce significant measurements of factors such as humidity, soil and air temperature, height of the crop, wind conditions, plant diameter and width and more in agricultural applications. Unmanned Aerial Vehicles (UAVs) which are also known as drones, Global positioning satellites and other data-gathering aircraft such as balloons and blimps usually have remote sensing equipment mounted.

Figure 4.5 illustrates this. Plants (C) receive electromagnetic energy (B) from the sun (A). Portion of the Electromagnetic energy is transmitted via the leaves. The reflected energy (D) is detected by the satellite's sensor. The information is then sent to the ground station (E). The information is then analysed (F) and plotted on field maps (G). Ground sensors, aerial sensors and satellite sensors are the three types of remote sensors that can be classified based on their enabling technologies. Ground sensors are available in a variety of configurations, including handheld, mounted on tractors and combines and free-standing in the field. Other common applications

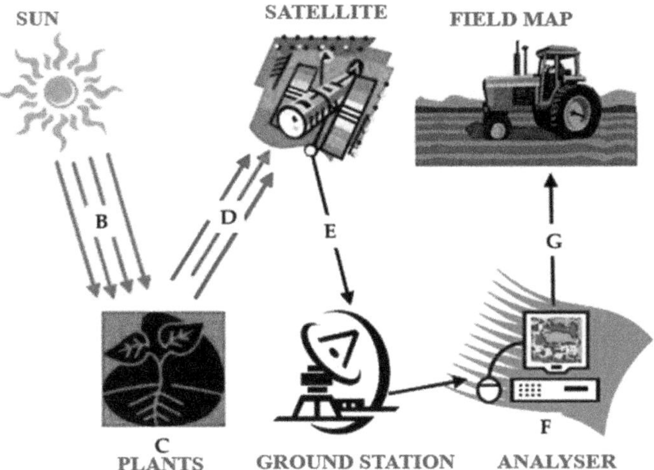


**Fig. 4.5** Illustrates a satellite remote sensing process as applied to agricultural monitoring processes

include determining nutrient levels in preparation for specific chemical and nutrient applications, evaluating weather and determining the moisture content of the soil.

### *4.5.2 Geographical Information System (GIS)*

A geographic information system (GIS) is a computer software database system that allows users to input, store, retrieve, analyse and view spatially referenced geographic data in a map like format. It is a computer-based management system that is used to calculate, store, analyse and display spatial data as a map. The GIS is the most important tool for extracting insights from data on variability. It is appropriately referred to as the "core" of precision farming.

It helps in the PA in two different ways:

- One is in the link and integration of GIS data (weather, crop, soil and field history, etc.) with simulation models.
- The next one is to support the engineering component for GPS guided machines and design implementation.

### *4.5.3 Global Positioning System (GPS)*

The Global Positioning System (GPS) is a constellation of satellites that can locate farm machinery's exact location in the field to within a metre. It is a critical technology for achieving accuracy when driving in the field, providing navigation and positioning

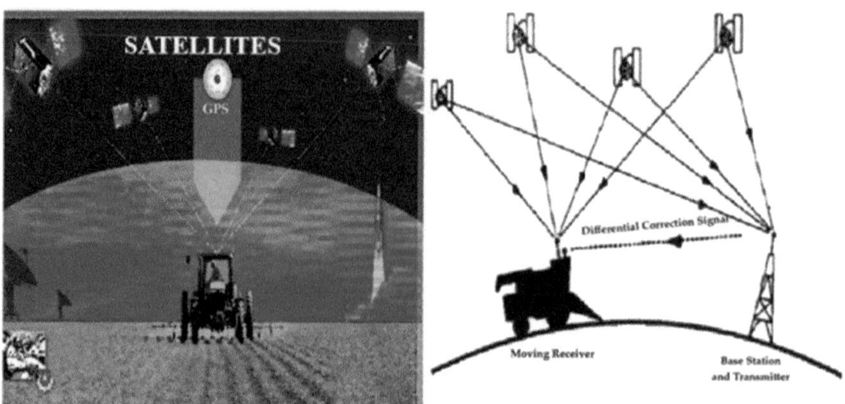


**Fig. 4.6** Global Positioning Systems (GPS)

expertise anywhere on the planet, at any time and in any condition. The systems use geographic coordinates (latitude and longitude) to record the location of the field and to locate and navigate agricultural vehicles inside a field with a 2-cm precision (Fig. 4.6).

#### 4.5.3.1 GPS Use in Agriculture

- **Tractor Guidance**
  Tractors cannot be automated. But with a recording GPS system if they plough their fields, then the tractor can then be programmed to follow the same route for cultivating, fertilizing, cultivating, harvesting and pest control. Farmers have potential to save a lot of money when the routes of tractors are programmed.

- **Soil Sampling**
  GPS and mapping software can be used to organize the collection of soil samples across a large field. In the field, sample areas can be way pointed and those coordinates are recorded in the mapping software. If the laboratory results are returned, the information can be plotted on maps and decisions about soil treatment can be made for different parts of the area. It is possible to save money and time by having variable rate applications and treating only certain regions.

- **Yield Monitoring**
  GPS allows for the assessment of yield variations across the farm. This is accomplished by dividing the area into zones and evaluating and plotting the yield of each zone on a map. There is another useful thing to be gained from the map, it can be used to obtain a better understanding of the region and to help direct future planting decisions.

- **Tracking Livestock**
  By attaching transmitters to the collars of valuable animals on a large farm, GPS can track the location of valuable animals. GPS transmitters can also be used to track the animals' location once they are directed to market.

## 4.5.4 *Role of Nano Technology in Precision Farming*

Nanotechnology is a multidisciplinary field of research. Efforts have recently been made to increase agricultural yield through extensive nanotechnology research. The green revolution ended in the indiscriminate use of pesticides and chemical fertilizers, resulting in the loss of soil biodiversity and the development of pathogen and pest resistance. Only nanoparticles or nano chips enable the delivery of materials to plants and the development of advanced biosensors for precision farming. Nano encapsulated herbicides, pesticides and conventional fertilizers helps in sustained and gradual release of nutrients and agrochemicals resulting in accurate dosage to the plants.

### 4.5.4.1 Delivery of Fertilizers by Nano Technology in Precision Agriculture

A large quantity of fertilizer is continuously used in the agriculture field to reasonably enhance food production. In precision farming, the right amount of fertilizers only used in the crop fields. Ammonium salts, nitrate, phosphate and urea are consumed as nutrients on the farm. In India, the large utilization of nitrogen fertilizer in the manner of urea enhances the agricultural production rate after the green revolution era. This increases atmospheric nitrous oxide level, which leads to higher atmospheric temperature and further base for global warming. To meet the N, P and K shortage level of soil, the known chemical fertilizers like urea, diammonium are applied. The fertilizers are not available to plants and Surplus quantity in the farm field have many harmful side effects and leads to pollution.

Nanomaterial coated fertilizers have a positive contribution to the gradual release of nutrients to plant growth. Since the surface tension of nanoparticles is higher than the conventional one, the recent progressive method for the sustainable release of fertilizer is effectively absorbed by plant or crop roots. Numerous synthetic and natural polymers have been utilized for this slow release of nutrients to plants. Of these, biodegradable polymers have excellent result in the sustainable release of NPK fertilizer to compensate for the nutrient losses and reducing the interaction of fertilizer with micro-organisms and water. Thus, nanotechnology is used in the PA fields for the proper usage of fertilizers.

#### 4.5.4.2 Solution for Desiccation in Bio-fertilizers by Nano Particles

The beneficial living microorganism is said to be bio fertilizers. Fungal mycorrhizae, Azotobacter, Rhizobium, blue green algae, Azospirillum, phosphate solubilizing bacteria such as Bacillus and Pseudomonas. In bio fertilizers, the microorganisms transform organic content into essential nutrient compounds suitable for improving soil fertility, keeping natural soil habitation.

Due to this advantage of microorganisms, the crop yield is enhanced. Organic farming is the primary goal of precision farming for reducing the health hazard of pesticides and fertilizers. Limitations of bio fertilizers are temperature sensitivity, minimum shelf life and desiccation problem. These constraints are controlled by nanotechnology. The polymer nanoparticle is one of the potential methods for coating bio fertilizers to resist quick moisture removal.

Water in oil emulsion also the better solution for desiccation limitations. The oil traps the water around microorganism; hence water evaporation is reduced. But the sedimentation is the more considerable disadvantage of this method. This limitation is resolved by hydrophobic silica nanoparticles, which minimized cell sedimentation and enhanced cell viability by thickening the oil phase throughout storage.

#### 4.5.4.3 Usage of Nano Bio-sensors in Precision Agriculture

In the precision farm, high-level technologies are used to identify the environmental variables, the nature of fertilizers, herbicides and pesticides to attain enhanced productivity from crops. The sensors, GPS and computers are utilized for the prediction of crop growth, yield, locality problems, soil pH, humidity, insects, weeds, etc., for accurate information of the fields. Nano biosensors assisted sustainable agriculture with all types of information in the fields. Nano sensors also detect plant viruses, soil nutrient levels and pathogens. For example, the agricultural pollutants can be monitored with the aid of carbon nanotube sensors with metal oxide nanoparticles.

## 4.6 Real-Time Applications and Instruments Used in Precision Farming

### *4.6.1 Use of Robotics in Precision Farming*

#### 4.6.1.1 Tilling

Due to the continuous growth of our population, there is a necessity of implementing the ultimate higher technologies in the farm fields to produce more food more sustainably in the same amount of agriculture field. Change in the face of traditional agriculture give benefits to all. The automatic tilling of the farm is a high

Fig. 4.7 Automatic tractor for tilling

necessity in precision farming. The concept of automatic tilling of agriculture fields by using above 400 hp engine is the trustworthy technology for very larger farm areas. The farmers can control and monitor the tilling operations in the associated mobile, tablet or desktop software. The driverless tilling machine can carry out its task, monitored remotely through the higher-end technology interface. Figure 4.7 shows the Automatic Tractor used for tilling.

#### 4.6.1.2 Harvesting

Selective harvesting is a promising technology in high-value crop fields. Currently, it is done by humans, which is one of the highly expensive and most labour-intensive agriculture works (Kootstra et al. 2021). But in precision farming required cost-effective methods for higher yield with minimum running cost. Many different types of robots were specifically developed for the selective harvest of vegetables and fruits. The researchers focused on the development of selective harvesting robotics technology. Robotic end-effectors were customized with a robotic knife to perform cutting action of crops while harvesting. In open fields, the selective harvest process is success and the speed are high due to the less complex structure of the field. Though selective harvesting technique adopted in precision farming, there is always a challenging task for an automated robot in the farm fields due to high levels of data discrepancy and incomplete information.

**Fig. 4.8** Harvesting robot

Use of robot in agricultural farms have the key issue such as higher capital cost of investment. The economic feasibility to adopt robots in all type of farming operations is minimum. Research selective harvesting technique adopted in precision farming, there is always a challenging task for an automated robot in the farm fields due to high levels of data discrepancy and incomplete information. Use of robot in agricultural farms have the key issue such as higher capital cost of investment. The economic feasibility to adopt robots in all type of farming operations is minimum. Research is going to implement the automated machinery in precision farming increases. Figure 4.8 shows the Harvesting Robot used in Precision farming.

#### 4.6.1.3 Monitoring

Utilizing robotics in precision farming is essential to gather the correct information about crop growth, weeds and soil. In a large scale of agriculture, the statistical records are processed through AI or human surveillance and the correct equipment used in the agriculture fields to observe the information at the right time.

The robots are employed in the farms to give the correct dosage of herbicides to control the weeds. Thus, the automated types of equipment contribute their role to make correct treatments with the necessary documentation. The autonomous robots can remotely monitor the status of the farm, conduct preventive activities

**Fig. 4.9** Agricultural robot used for monitoring

and collect large data for the process. Thus, the robotic monitoring application meets the higher service quality requirement for farming activities. Generally, autonomous robot moves in the farm field in an autonomous mode, programmed not tot damage cultivated crops, wandering around the garden beds to gather the state of crop growth and soil properties, the presence of thieves and insects, mature fruit and variable parameters regarding the crop ultimately affect the harvest. On the computer itself, the farmer can collect the relevant data's and take necessary precaution measures, if needed. Figure 4.9 shows the Agriculture Robot setup used for monitoring the crops in the agricultural field.

#### 4.6.1.4 Herbicides Spraying by Robots

Weed control is essential in all types of agriculture systems. The weed is controlled by several methods such as mechanical, biological, cultural and chemical methods. For example, in mechanical weed control techniques, the primary tillage is done for burying the perennial seeds in depth. Due to this technique, the weed seeds are prevented from emerging with cultivated crops for a certain period. The use of herbicides in chemical means of weed control gave higher production efficiency, less mechanical tillage practice and efficient all types of weed control. But the use of enormous herbicides in agricultural fields leads to human health problems and contamination in the soil and water. Thus, each technique has several unique advantages with few drawbacks itself.

Precision farming techniques can be used in weed management by utilizing all advantages of different weed control techniques for attaining the key benefits such as reducing cost, saving time and taking care of environment. Reduce usage of

**Fig. 4.10** Robotic weed killer

herbicides by applying variable rates based on the nature of the weed patches as a replacement for using herbicides uniformly throughout the field and implementing new methodologies such as global positioning systems to identify the denser weeded surfaces. The weed mapping by robotic applications brings the data's about weed population and distribution in the fields. Precision farming techniques with robots gave many benefits while using weed management techniques. Automation replaces the human labour dependence to perform several operations across the fields. Generally, in farm fields, the manual labour work having repetitive tasks. The main objective of utilizing robotic tools is to minimize the dependency of labour around the clock and the other side enhances the efficiency, crop production quantity and quality. The combination of sensors and GPS facilities empowers the automation process irrespective of external factors such as wind speed, topography and geological conditions. The effective herbicides spraying is done on the requirement point based on the mined data from fields. Thus, the wastage is lowered. Figure 4.10 shows the Robotic Weed Killer used in the Precision agriculture.

## *4.6.2 Usage of Drones in Precision Farming*

### 4.6.2.1 Fertilizer Spraying

The proper dosage of fertilizer at the right moment and in the right way typically implies good agriculture practices. The correct dosage means it depends on the soil or crop condition; the nutrients can supply. The soil sampling is done to analyse the nutrient amount available in the soil. The quick method of soil sampling is

**Fig. 4.11** Fertilizer spraying using drone

made by using optical techniques or electrochemical sensors. For large scale farming practices, the drones help to meet the nutrient requirement of crops without manual human intervention. Manure or organic fertilizer is applied into the soil before crop planting. The Drones supply the fertilizer on the crop fields. The sensors in the drone measure the fertilizer mass flow rate of granular or liquid nutrient fertilizer. The information from the Global Positioning System can be utilized to guide the drone for the specific task of supplying the correct amount of nutrients to agriculture fields. Figure 4.11 shows the Fertilizer sprayer by using the Drone in agriculture field.

#### 4.6.2.2 Surveillance

An agriculture drone is an unmanned aerial vehicle. Drones might be used for monitoring crop health, detecting crop disease and spraying pesticides and herbicides on crops. Now in recent research, drones are worked in swarms. Latest agriculture drones connected with internet connectivity for remote access and also communicate

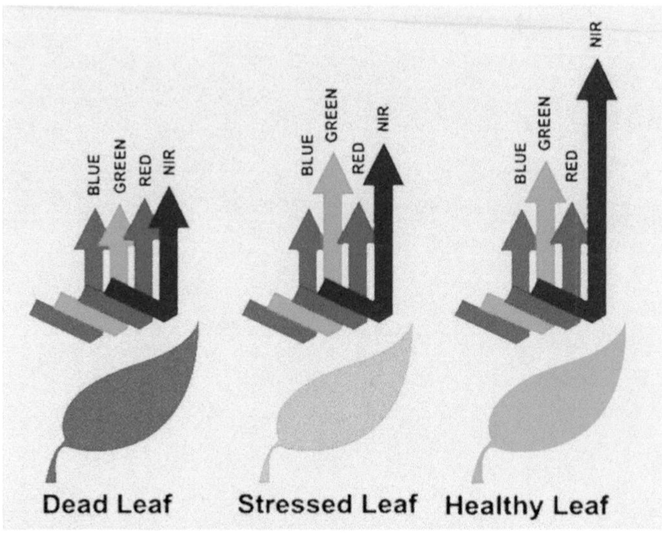


**Fig. 4.12** Plant health monitoring using drone

with GPS signals to trace the specific site location. Surveillance provides information about weed, crop health, animals' intrusion and crop safety.

#### 4.6.2.3 Yield Monitor

The crop yield should be monitored continuously in each agriculture field. The data is displayed on the map as output with the help of Geographic information system software solutions. Thus, yield maps gave the crop production data for deciding the specific crop management. Otherwise, the uniform management may be applicable across the field inarticulate to the field property. The yield monitors provide information on past crop management, topography and nutrient usage for future investment in the specific crop fields. The drones are used for the yield monitor application, offers the opportunity to reduce the efforts of manual monitoring work, which replaces another labour-demanding task. The yield monitor also sends the specific information to the robot for exact harvest locations. The different type of information retrieved using drone images in the web-based application used by the observer is shown in Fig. 4.12. Based upon the information retrieved the farmer/observer can take decisions accordingly.

## 4.7 Benefits of Precision Farming

The important benefits of precision farming are shown in Fig. 4.13.

**Fig. 4.13** Benefits of precision farming

Usually, three crucial components of precision farming are information, technology and management. The ultimate gains of modern precision farming are,

- Better quality of crop yield.
- Utilize all advantages of various farming techniques.
- More cost-effective farming process.
- Low wastage and manual or human effort utilization
- Increase profit by simplified farming Practices.
- Producing more with lesser effort.
- Provide optimal conditions of plant growth by the continuous monitor of soil and crop parameters.
- Regular updates of the crop field to do specific precaution actions.
- Give precise data for management decisions.
- Save cost and time, reduce fertilizer, pesticides and herbicides usage because of specific treatment on a specific location.
- Maintain past farm record for future investment.
- Integrated with farm management software to diagnose the farm records.
- Reduce the misuse of pesticides due to the lack of awareness in farm field parameters.
- Remote access of sensors and automated machineries.

## 4.8 Conclusion

The various ideas and technologies that will make up tomorrow's PA are still in the early stages of development. The present work detailed about how the various emerging technologies like AI, IoT, GPS, GIS, Remote sensing used in Precision Farming. In addition to that, the present work discussed about the real-time applications of the above-mentioned technologies assisted robots and drones used in modern agriculture. Through the implementation of the emerging technologies in our traditional farming make it more precise and modern in terms of productivity, greater sustainability, economic benefits and environmental protection. In the upcoming future, incorporating technologies like block chain and similar technologies in the precision farming makes it superior.

## References

Adamchuk VI, Hummel JW, Morgan MT, Upadhyaya SK (2004) On-the-go soil sensors for precision agriculture. Comput Electron Agric 44(1):71–91

Al Bashish D, Braik M, Bani-Ahmad S (2011) Detection and classification of leaf diseases using K-means-based segmentation and Neural-networks-based classification. Inf Technol J 10(2):267–275

Al-Hiary H, Bani-Ahmad S, Reyalat M, Braik M, Alrahamneh Z (2011) Fast and accurate detection and classification of plant diseases. Int J Comput Appl 17(1):31–38

Ampatzidis Y, Partel V, Costa L (2020) Agroview: cloud-based application to process, analyze and visualize UAV-collected data for precision agriculture applications utilizing artificial intelligence. Comput Electron Agric174:105457

Arif C, Mizoguchi M, Setiawan BI (2013) Estimation of soil moisture in paddy field using Artificial Neural Networks. 1303.1868

Baggio A (2005) Wireless sensor networks in precision agriculture. In: ACM workshop on real-world wireless sensor networks (REALWSN 2005), Stockholm, Sweden 20, pp 1567–1576

Balleda K, Satyanvesh D, Sampath NVSSP, Varma KTN, Baruah PK (2014) Agpest: an efficient rule-based expert system to prevent pest diseases of rice & wheat crops. In: 2014 IEEE 8th international conference on intelligent systems and control (ISCO). IEEE, pp 262–268

Banerjee G, Sarkar U, Ghosh I (2017) A radial basis function network based classifier for tea pest detection. IJARCSSE 7(5):665–669

Bannerjee G, Sarkar U, Das S, Ghosh I (2018) Artificial intelligence in agriculture: a literature survey. Int J Sci Res Comput Sci Appl Manag Stud 7(3):1–6

Barrero O, Rojas D, Gonzalez C, Perdomo S (2016) Weed detection in rice fields using aerial images and neural networks. In: 2016 XXI symposium on signal processing, images and artificial vision (STSIVA). IEEE, pp 1–4

Barros M, Fernandes R (2015) An approach via fuzzy systems for dynamics and control of the soybean aphid. In: Proceedings of IFSA-EUSFLAT

Beznosov AG, Skvortsov EA, Skvortsova EG (2021) Prospects for application of blockchain technology in land administration. In: IOP conference series: earth and environmental science. IOP Publishing, 699(1), 012045

Boulanger AG (1983) The expert system PLANT/CD: a case study in applying the general purpose inference system ADVISE to predicting black cutworm damage in corn. PhD thesis, University of Illinois at Urbana-Champaign

Boyd DW, Sun MK (1994) Prototyping an expert system for diagnosis of potato diseases. Comput Electron Agric 10(3):259–267

Bralts VF, Driscoll MA, Shayya WH, Cao L (1993) An expert system for the hydraulic analysis of microirrigation systems. Comput Electron Agric 9(4):275–287

Broner I, Comstock CR (1997) Combining expert systems and neural networks for learning site-specific conditions. Comput Electron Agric 19(1):37–53

Burks TF, Shearer SA, Gates RS, Donohue KD (2000) Back propagation neural network design and evaluation for classifying weed species using color image texture. Trans ASAE 43(4):1029

Burks TF, Shearer SA, Heath JR, Donohue KD (2005) Evaluation of neural-network classifiers for weed species discrimination. Biosys Eng 91(3):293–304

Capizzi G, Lo Sciuto GRAZIA, Napoli C, Tramontana E, Woźniak M (2016) A novel neural networks-based texture image processing algorithm for orange defects classification. Int J Comput Sci Appl 13(2)

Costa L, Archer L, Ampatzidis Y, Casteluci L, Caurin GA, Albrecht U (2020) Determining leaf stomatal properties in citrus trees utilizing machine vision and artificial intelligence. Precis Agric: 1–13

Cox S (2002) Information technology: the global key to precision agriculture and sustainability. Comput Electron Agric 36(2–3):93–111

Dahikar SS, Rode SV (2014) Agricultural crop yield prediction using artificial neural network approach. Int J Innov Res Elect Electron Instrumen Control Eng 2(1):683–686

Duhan JS, Kumar R, Kumar N, Kaur P, Nehra K, Duhan S (2017) Nanotechnology: The new perspective in precision agriculture. Biotechnol Rep 15:11–23

Eddy PR, Smith AM, Hill BD, Peddle DR, Coburn CA, Blackshaw RE (2008) Hybrid segmentation–artificial neural network classification of high resolution hyperspectral imagery for site-specific herbicide management in agriculture. Photogramm Eng Remote Sens 74(10):1249–1257

Escobar C, Galindo J (2004) Fuzzy control in agriculture: simulation software. In: Industrial simulation conferences, pp 45–49

García L, Parra L, Jimenez JM, Lloret J, Lorenz P (2020) IoT-based smart irrigation systems: an overview on the recent trends on sensors and IoT systems for irrigation in precision agriculture. Sensors 20(4):1042

Ghosh I, Samanta RK (2003) TEAPEST: an expert system for insect pest management in tea. Appl Eng Agric 19(5):619

Gottschalk K, Nagy L, Farkas I (2003) Improved climate control for potato stores by fuzzy controllers. Comput Electron Agric 40(1–3):127–140

Hu S, Huang S, Huang J, Su J (2021) Blockchain and edge computing technology enabling organic agricultural supply chain: a framework solution to trust crisis. Comput Ind Eng153:107079

Huang KY (2007) Application of artificial neural network for detecting Phalaenopsis seedling diseases using color and texture features. Comput Electron Agric 57(1):3–11

Jesus J, Panagopoulos T, Neves A (2008) Fuzzy logic and geographic information systems for pest control in olive culture. In: Proceedings of of the 4th WSEAS international conference on energy, environment, ecosystems & sustainable development

Kashyap PK, Kumar S, Jaiswal A, Prasad M, Gandomi AH (2021) Towards precision agriculture: iot-enabled intelligent irrigation systems using deep learning neural network. IEEE Sens J

Kavdir I, Guyer DE (2004) Apple grading using fuzzy logic. Turk J Agric for 27(6):375–382

Keswani B, Mohapatra AG, Keswani P, Khanna A, Gupta D, Rodrigues J (2020) Improving weather dependent zone specific irrigation control scheme in IoT and big data enabled self driven precision agriculture mechanism. Enterp Inf Syst 14(9–10):1494–1515

Khan FS, Razzaq S, Irfan K, Maqbool F, Farid A, Illahi I, & Amin TU (2008) Dr. Wheat: a web-based expert system for diagnosis of diseases and pests in Pakistani wheat. In: Proceedings of the world congress on engineering, 1, 2–4

Khoshnevisan B, Rafiee S, Omid M, Yousefi M, Movahedi M (2013) Modeling of energy consumption and GHG (greenhouse gas) emissions in wheat production in Esfahan province of Iran using artificial neural networks. Energy 52:333–338

Kootstra G, Wang X, Blok PM, Hemming J, Van Henten E (2021) Selective harvesting robotics: current research, trends, and future directions. Curr Robot Rep; 1–10

Lee J, Nazki H, Baek J, Hong Y, Lee M (2020) Artificial intelligence approach for tomato detection and mass estimation in precision agriculture. Sustainability 12(21):9138

Li SK, Suo XM, Bai ZY, Qi ZL, Liu XH, Gao SJ, Zhao SN (2002) The machine recognition for population feature of wheat images based on BP neural network. Agric Sci China 1(8):885–889

Liaghat S, Balasundram SK (2010) A review: the role of remote sensing in precision agriculture. Am J Agric Biol Sci 5(1):50–55

Liu G, Yang X, Li M (2005) An artificial neural network model for crop yield responding to soil parameters. In: Proceedings of international symposium on neural networks. Springer, Berlin, Heidelberg

Liu W, Shao XF, Wu CH, Qiao P (2021) A systematic literature review on applications of information and communication technologies and blockchain technologies for precision agriculture development. J Clean Prod 126763

Nabavi-Pelesaraei A, Abdi R, Rafiee S (2016) Neural network modeling of energy use and greenhouse gas emissions of watermelon production systems. J Saudi Soc Agric Sci 15(1):38–47

Nakano K (1997) Application of neural networks to the color grading of apples. Comput Electron Agric 18(2):105–116

Nebot P, Torres-Sospedra J, Recatala G (2012) Using neural networks for maintenance tasks in agriculture: precise weed detection

Pahlavan R, Omid M, Akram A (2012) Energy input–output analysis and application of artificial neural networks for predicting greenhouse basil production. Energy 37(1):171–176

Pasqual GM (1994) Development of an expert system for the identification and control of weeds in wheat, triticale, barley and oat crops. Comput Electron Agric 10(2):117–134

Pierce FJ, Nowak P (1999) Aspects of precision agriculture. Adv Agron 67:1–85

Ponnusamy V, Natarajan S (2021) Precision agriculture using advanced technology of IoT, unmanned aerial vehicle, augmented reality, and machine learning. In: Smart sensors for industrial Internet of Things. Springer, Cham, pp 207–229

Prakash C, Rathor AS, Thakur GSM (2013) Fuzzy based Agriculture expert system for Soyabean. In: Proceedings of international conference on computing sciences WILKES100-ICCS2013, Jalandhar, Punjab, India

Raj EFI, Balaji M (2021) Analysis and classification of faults in switched reluctance motors using deep learning neural networks. Arab J Sci Eng 46(2):1313–1332

Robert PC (2005) Precision agriculture: a challenge for crop nutrition management. In: Progress in plant nutrition: plenary lectures of the XIV international plant nutrition colloquium. Springer, Dordrecht, pp 143–149

Robinson C, Mort N (1997) A neural network system for the protection of citrus crops from frost damage. Comput Electron Agric 16(3):177–187

Roussel O, Cavelier A, van der Werf H (2000) Adaptation and use of a fuzzy expert system to assess the environmental effect of pesticides applied to field crops. Agr Ecosyst Environ 80(1):143–158

Saini HS, Kamal R, Sharma AN (2002) Web based fuzzy expert system for integrated pest management in soybean. Int J Inf Technol 8(1):55–74

Samanta RK, Ghosh I (2012) Tea insect pests classification based on artificial neural networks. Int J Comput Eng Sci 2(6):1–13

Sanjeevi P, Prasanna S, Siva Kumar B, Gunasekaran G, Alagiri I, Vijay Anand R (2020) Precision agriculture and farming using Internet of Things based on wireless sensor network Trans Emerg Telecommun Technol 31(12):e3978

Sannakki SS, Rajpurohit VS, Nargund VB, Kumar A, Yallur PS (2011) Leaf disease grading by machine vision and fuzzy logic. Int J Comp Tech Appl 2(5):1709–1716

Sarma SK, Singh KR, Singh A (2010) An expert system for diagnosis of diseases in rice plant. Int J Artif Intell 1(1):26–31

Shi Y, Yuan H, Liang A, Zhang C (2007) Analysis and testing of weed real-time identification based on neural network. In: Proceedings of international conference on computer and computing technologies in agriculture. Springer, Boston, MA, pp 1095–1101

Shi Y, Zhang C, Liang A, Yuan H (2007) Fuzzy control of the spraying medicine control system. In: Proceedings of international conference on computer and computing technologies in agriculture. Springer, Boston, MA

Si Y, Liu G, Lin J, Lv Q, Juan F (2007) Design of control system of laser leveling machine based on fuzzy control theory. In: Proceedings of international conference on computer and computing technologies in agriculture. Springer, Boston, MA, pp 1121–1127

Sicat RS, Carranza EJM, Nidumolu UB (2005) Fuzzy modeling of farmers' knowledge for land suitability classification. Agric Syst 83(1):49–75

Siraj F, Arbaiy N (2006) Integrated pest management system using fuzzy expert system. In: Proceedings of KMICE-2006, University of Malayasia, Sintok

Soheili-Fard F (2015) Seyed Babak Salvatian, Forecasting of tea yield based on energy inputs using artificial neural networks (a case study: Guilan province of Iran). Biol Forum 7(1):1432–1438

Srinivasan A (ed) (2006) Handbook of precision agriculture: principles and applications. CRC Press

Stafford JV (2000) Implementing precision agriculture in the 21st century. J Agric Eng Res 76(3):267–275

Sumarudin A, Ismantohadi E, Puspaningrum A, Maulana S, Nadi M (2021) Implementation irrigation system using support vector machine for precision agriculture based on IoT. In: IOP conference series: materials science and engineering. IOP Publishing, 1098(3), 032098

Taki M, Ajabshirchi Y, Ranjbar SF, Matloobi M (2016) Application of neural networks and multiple regression models in greenhouse climate estimation. Agric Eng Int CIGR J 18(3):29–43

Talaviya T, Shah D, Patel N, Yagnik H, Shah M (2020) Implementation of artificial intelligence in agriculture for optimisation of irrigation and application of pesticides and herbicides. Artif Intell Agric

Tilva V, Patel J, Bhatt C (2013) Weather based plant diseases forecasting using fuzzy logic. In: Proceedings of (NUiCONE). IEEE

Tremblay N, Bouroubi MY, Panneton B, Guillaume S, Vigneault P, Bélec C (2010) Development and validation of fuzzy logic inference to determine optimum rates of N for corn on the basis of field and crop features. Precis Agric 11(6):621–635

Valdés-Vela M, Abrisqueta I, Conejero W, Vera J, Ruiz-Sánchez MC (2015) Soft computing applied to stem water potential estimation: a fuzzy rule based approach. Comput Electron Agric 115:150–160

van der Werf H, Zimmer C (1998) An indicator of pesticide environmental impact based on a fuzzy expert system. Chemosphere 36(10):2225–2249

Wei MCF, Maldaner LF, Ottoni PMN, Molin JP (2020) Carrot yield mapping: a precision agriculture approach based on machine learning. AI 1(2):229–241

Yang Q (1993) Classification of apple surface features using machine vision and neural networks. Comput Electron Agric 9(1):1–12

Zhang N, Wang M, Wang N (2002) Precision agriculture—a worldwide overview. Comput Electron Agric 36(2–3):113–132

# Chapter 5
# ML-Based Smart Farming Using LSTM

**Himadri Nath Saha and Reek Roy**

**Abstract** There has been a noticeable change in agricultural procedures from conventional techniques in the current era. The Internet of Things (IoT) and Machine Learning (ML), along with many other significant technical advancements in recent decades, are game-changing fields with the ability to transform agricultural industries. Recent technological improvements have had a major effect on agriculture, and it has been proven that IoT can be utilized in agriculture to increase agricultural efficiency. The need to modernize the agriculture sector has grown as the demand for food has risen in both quantity and quality. Scientists and researchers are increasingly interested in applying ML and the IoT to the field of agriculture. Sensors and the IoT are critical for pushing the world's agriculture towards a more prosperous and sustainable future. It aids farmers in the quality of their farmland in order to meet the world's food demand. We have proposed a system for increasing crop yields that includes a tracking model that uses Wireless Sensor Networks (WSN) and cloud computing to manage parameters and gather information about the field's soil using temperature sensors, moisture sensors, air humidity sensors, pH sensors, and RGB-D sensors. The data collected by the sensors will be sent to a cloud server for analysis. Following the reception of those values, the server can take specific decisions depending on the analysis method. A mobile phone application along with a website can be built to keep track of the real-time data. The sensor devices that are installed in the field may get damaged or stop working, or some other issue might crop up in our system, like the security of a sensor is damaged. In such scenarios, we aim to make our smart farming system more intelligent by incorporating a Machine Learning algorithm in our framework like the Long Short-Term Memory (LSTM) network. The LSTM will be able to detect any anomalies of the environmental parameters and the environmental parameters of the next moment can be predicted by studying the agricultural climate parameters of the current time in order to accomplish the goal of early alert; for example, if the temperature sensor reads a wrong temperature due to

H. N. Saha (✉)
Department of Computer Science, Surendranath Evening College, Calcutta University, Kolkata, West Bengal, India

R. Roy
Department of Computer Science, Belda College, Vidyasagar University, Paschim Medinipur, Belda, West Bengal, India



A. Choudhury et al. (eds.), *Smart Agriculture Automation using Advanced Technologies*, Transactions on Computer Systems and Networks,
https://doi.org/10.1007/978-981-16-6124-2_5

some issue, then the LSTM model will be able to predict the approximately correct temperature required. The IoT system will work both parallelly and simultaneously with the existing hardware systems, along with the incorporation of ML. This will lead to an effective crop management solution.

**Keywords** Machine learning (ML) · Long short-term memory (LSTM) · Internet of things (IoT) · Wireless sensor networks (WSN) · Cloud computing · Smart agriculture

## 5.1 Introduction

In order to minimize inefficiencies and increase efficiency across all markets, the Internet of Things (IoT) is starting to reach a wide range of sectors and industries, ranging from manufacturing, healthcare, networking, and power to agriculture and farming. In the last several decades, machine, Internet, and mobile Internet technologies have brought many significant improvements to human society. IoT modules include Wireless Moisture Sensor Networks (WMSN) and Wireless Sensor Networks (WSN). Throughout humanity's civilization, significant developments have been made to increase agricultural output with less capital and labour efforts. Despite this, the high population density has never enabled demand and supply to be balanced overall these periods (Ayaz et al. 2019). The Internet of Things (IoT) is a web of sensors and networking that enables applications such as maximum irrigation in farming. The Wireless Sensor Network (WSN) can be used to solve a wide range of problems. Monitoring of health, agriculture, temperature, air quality, and ground slide control are all examples of WSN. Networking technology, sensor technology, information processing technology, control technology, and database storage are all part of the WSN (Mat et al. 2016).

The Internet of Things integrates computing devices embedded in ordinary objects, enabling them to transmit and receive data. There seem to be two benefits: one can enable the machines to gather knowledge about the environment all without relying on people and one can also minimize indulgence, waste, and expense by analysing the data gathered. The Internet of Things enables real and artificial worlds to connect. Actuators and sensors enable the digital and physical worlds to connect. These sensors gather data that needs to be collected and analysed. Collection of information may take place on a remote server, in the cloud, or at the network's edge (Garg and Dave 2019).

Food will always be in need all throughout the globe. With the rise of the global populace, there is a scarcity of food due to a complete lack of technological convergence and consumption. In simple terms, agriculture is the practice of planting and growing the soil, producing outputs and raising animals, and finally selling the resulting products. Combining technologies fostering practises will meet the growing demand for food in terms of quantity and quality. The global population is increasing, which necessitates an increase in crop production to meet the worldwide demand for food consumption and nutrition (Saha et al. 2021a).

In agriculture, the use of IoT in processes such as crop production, processing, storing, shipping, advertising, and retailing makes it much easier for farm owners and middlemen to complete their tasks efficiently and profitably. Several factors influence the production of a crop during cultivation, including fertility of the soil, crop irrigation, the pH content of the soil, the size of the leaf, etc. With the aid of IoT, these variables can be easily tracked. By keeping track of the harvest in the early stages and keeping an eye on the above factors, the yield of the crop can be increased. IoT could be used to ensure that no grain is lost during cultivation. The entire output is achieved efficiently and without flaws. Using IoT, the position of representatives and other intermediaries can be minimized to the bare minimum. IoT assists us in ensuring that the correct temperature and relative humidity are kept in the vehicles used for shipping the harvested crop. Radio Frequency Identification (RFID) tags can be used to track storage by efficiently gathering data about the available inventory stock in the warehouse which can be used to track the stock of the harvest in the facility. Using IoT software, it is possible to effectively regulate stock in retail stores. As a result, IoT occupies a crucial role in agriculture, from the initial phase of planting to the final process of selling. It aids in the reduction of the number of agents in the agricultural supply chain, resulting in cost savings for both producers and end-users (Jyothi and Nandan 2020).

Cloud computing provides a wide range of resources and strategies for processing, storing, and analysing the massive amounts of data generated by computers. These tools and methods can then be used to automate processes, forecast situations, and optimize a variety of real-time processes that enable farmers to make decisions about sustainable agriculture. Plant diseases, for example, are traditionally diagnosed by agriculture professionals manually, which can be unreliable and usually takes a lot of time on occasions. Controlled and automatic remote detection of agricultural crops diseases may be used for Wireless Sensor Networks (WSN) and Internet of Things (IoT) to prevent erroneous findings and evade time-consuming procedures (Saha et al. 2021b).

Machine Learning (ML) is a form of Artificial Intelligence (AI) that allows machinery to develop from their past experiences. Its algorithms use statistical techniques to derive specifically from databases rather than relying on a model of preset formulas. If the amount of training data available grows, the algorithms evolve to improve their accuracy. Machine learning algorithms are classified into two classifications: supervised and unsupervised learning. Supervised learning trains a system to determine the output variable for sample data using a known collection of labelled information. Unsupervised computing, on the other hand, uses hidden correlations or underlying constructs in data to make deductions from unlabeled data. It is beneficial for explorative purposes where there will be no definite target in mind or the detail in the dataset is not obvious. It is also great for dimension reduction on information with a lot of different functions (Mekonnen et al. 2019).

Not only will the incorporation of IoT along with Machine Learning, with agriculture benefits farmers, but it will also benefit the nation's growth. As a result, the farming industry will be transformed by the implementation of productive smart agricultural systems; which can be termed smart farming also. Agriculture and IoT, along with ML integration, assumes that links can be wirelessly categorized depending on power usage, uplink and downlink data rates, interface per success node, topography, channel width, size of the packet, and available frequency bandwidth. IoT-enabled agricultural systems aid in the discovery of precise information about atmospheric environments, soil conditions, and other factors that affect productivity (Saha et al. 2021b).

Technology is crucial in reducing the stresses that the agricultural industry faces as a consequence from factors such as growing population, consumer demands, and rising demand for land, water, electricity, and fuel, both of which are becoming increasingly scarce. Precision Agriculture (PA) is another term for smart farming, which is similar to other machine-to-machine (M2M)-based applications such as smart city and smart environment. One of several sectors where ML strategies are evolving to measure and interpret the big data within that area is smart farming allowed by WSN and IoT. In IoT smart data processing, both supervised and unsupervised learning approaches are widely used across different domains. Grain management, animal management, water treatment and management, and soil management are all examples of ML applications in Precision Agriculture (PA). Yield estimation, disease identification, weed discovery, and phenotype identification are all applications of machine learning in crop maintenance and management (Mekonnen et al. 2019).

With agriculture playing such an important role in a country's economic development, it is only natural to improve this sector using any of the new technology available. Sensor systems that capture information from the soil, harvest, different atmospheric characteristics, animal behaviour, and tractor condition should all be integrated into smart farming. These sensor data readings will provide the producer, that is, the farmer with relevant details on weather patterns and predictions, field tracking and harvest estimation, identification of plant and animal disease, and more by the use of IoT computing incorporated with ML and analysis.

This chapter of the book focuses on the creation of an IoT-enabled agricultural system using a Machine Learning (ML) algorithm, as well as the implementation of a system with modern technology that will support farmers and improve market profit, efficiency, and product quality. To overcome the difficulties faced by the farmers like manual soil monitoring, crop health monitoring, maintaining the soil's moisture manually, this system has been built which uses an array of sensors to measure and store the environmental conditions along with the soil conditions. The ML algorithm has been used to predict temperature and humidity for the future.

Section 5.2 consists of a literature survey done on various existing systems mentioning the results, pros, and cons of these systems. The proposed model's device design and architecture have been detailed in Sect. 5.3. The methodology of the proposed model is explained in Sect. 5.4. The performance review and future research strategy of the proposed model have been explained in Sects. 5.5 and 5.6. This book chapter's conclusion is detailed in Sect. 5.7.

## 5.2 Background and Related Works

When harvesting crops, farmers must keep a close eye on a multitude of elements in the fields. The crops must be checked on a regular basis. Agriculture requires the farmers to pay minute attention to changes in soil quality, such as pH changes, moisture present in the soil, and how many macronutrients such as nitrogen, phosphorus, potassium, etc., is required, in order to maximize agricultural productivity (Saha et al. 2021a). Farmers have typically and traditionally kept an eye on their crops manually by going to the fields and monitoring the crops. Professionals, for instance, have determined which crops demand which and how much fertilizer; and which pesticides for pest control. As a result, we can cater to see that there is a requirement for IoT-based smart surveillance and surveillance systems in the farming sector. Climate conditions such as intense weather, hurricanes, floods, and hail, among others, affect crop development. Farmers must deal with changes in environmental patterns, such as precipitation, rain, day and night temperatures, wind, and so on, on a constant basis (Saha et al. 2021a). Predictive analytics is most often used in conjunction with satellite measurements to forecast climate and agricultural survival, as well as in pest and disease detection and remote precision agriculture applications. For future modelling and decision models, predictive analytics is used in the data collection, managing, and interpretation of sensor information. ML approaches are also widely used as decision support in IoT WSN-based irrigation systems (Mekonnen et al. 2019).

Diseases in plants are a significant threat to agriculture. These diseases are traditionally identified by scientists manually, which happens to be not only time-consuming but also unreliable at places. Plant infections may well be identified with the aid of WSN and IoT. Crop disease prediction and soil quality control are two important fields where machine learning approaches have been used for quite some time. In many places, Plantix, which is a tool for image recognition is used; it utilizes machine learning methods to identify soil abnormalities and plant infections in agriculture using soil variations in its software algorithm (Mekonnen et al. 2019). The use of WSN and IoT in agriculture, including a thorough analysis of detectors and IoT information analysis for agriculture applications, using Machine Learning (ML) techniques will help us give a very good framework that will benefit the farmers a lot.

Saha et al. (2021a) proposed a design to enhance the agricultural process. The proposed framework uses a Raspberry Pi board and a variety of sensors like a pH sensor, a capacitance dielectric soil moisture sensor, and a Passive Infrared Sensor (PIR sensor) attached to it. The farmer will be told to take action if the soil nutrient, pH, or moisture levels are not up to par with the specifications for a given crop through a buzzer and LED. The system also includes a tracking system for checking crop health and avoiding pest control. This monitoring system has an Unmanned Aerial Vehicle (UAV) equipped with an RGB-D sensor and also keeps in check the crops' health. Sensor data is gathered and processed in the cloud, and the farmer is alerted by a phone app that utilizes the Global System for Mobile Communications (GSM). A website is also present which shows the status of the sensor readings.

The same readings are reflected in the phone application. The buzzer and LEDs are used to mark the threshold values. The buzzer switches on and the LED lights up if the sensor readings cross the threshold value for a specific crop. This system can be further developed by integrating an ML model so that detection of anomalies of sensor readings can be done and an early alert can be sent to the user.

Mekonnen et al. (2019) provides a thorough comparative study of the use of various machine learning algorithms in sensor big data analysis in the agroecosystem. Along with that, a case study on a distributed WSN has been presented. The WSN has been built using sensor nodes connected to an Arduino micro-controller. Zigbee wireless connection has been used to collect, study as well as control the parameters of the soil, environment, and weather for the growth of crops in critical situations.

Varman et al. (2017) pursued two key goals in their study: one is forecasting the best crop for the following harvest cycle and the other is improving the field's irrigation infrastructure by a process of selective irrigation. The proposed work comprises three important units: (1) Remote WSN (2) Cloud (3) Twilio and mobile Application. The above target is accomplished by tracking the field on a regular basis. The inspection method entails gathering knowledge about the field and its soil properties. A ZigBee and GSM-based WSN was set up to gather this information and to take a look back at it by uploading it to the cloud on a regular basis. Seven nodes in the WSN were used to calculate the temperature of the soil, ambient temperature, and levels of humidity in the sample region using the following sensors: LM35, Soil moisture sensor, DS18B20, and DHT11. The data was compiled and stored on a five-minute frequency basis. The sensor readings were recorded for a period of 12 months, and this vast amount of data was stored in the database. The peak and lowest values for each parameter were calculated which was used to develop the ML model. Three different methods were used in the learning-based classification system: (1) Feedforward, (2) Long Short-Term Memory (LSTM), and (3) Gated Recurrent Unit (GRU) on the collection of data, with the precision of the results confirmed by the training dataset. The analytics are based on the data that has been stored. LSTM networks were discovered to be the most appropriate algorithm after extensive testing. This system can be further developed using other sensors to measure other factors of the surrounding or a drone can be used to keep constant monitoring. The expected results are then compared to the ideal parameters and the owner is notified of the best-suited harvest for the next year via SMS.

Joshi et al. (2017) proposed a framework that used Bayesian machine learning approaches to predict output parameters accurately using a probability distribution. This research suggested IoT-based small-scale farming as a viable option. A person will be able to maintain his own farm in small and private gardens by using the developed framework. The user can just choose which plants to farm in his garden or backyard using a basic web application and the rest can be taken care of by the bot. This bot in the proposed approach is able to sow the seeds, water every plant at the appropriate times, and even is able to plot the weed that is to be buried. For each type of plant, a dataset is supplied updated in a database with details on many criteria that must be monitored. Different sensors (light intensity, temperature, and humidity sensors) are used to detect soil and environmental attributes that can be used to predict impending changes and eventually take appropriate measures. The majority of the hardware used is inexpensive and easy to obtain. The individual is presented with an integrated software application that allows him to conveniently customize and monitor the Farm bot using a web browser on their computers, or other technological smart gadgets. The webpage accepts the user's input commands and allows him to plan the farm or the garden. Both communications seen between the web service and the Farm bot devices are routed via the cloud service Message Queuing Telemetry Transport (MQTT) gateway. The Raspberry Pi controller was used to connect with the Arduino over USB, as well as send and receive data. Image processing has been shown to be effective in preventing weed development, as evidenced by the results in this model. The suggested simple approach for recognizing shadow or moistened ground as backdrop appears to be successful, eliminating the need for complicated extraction strategies for the foreground. The proposed model can be further developed for large-scale farming and other sensors can be attached to enhance the system. The sensors collect information regarding a variety of factors like the atmospheric pressure, the pH level of the soil, the temperature, direction of the wind, etc. All the sensor data are stored in a local MySQL database through an IoT gateway. The router is based on mesh topology and is Linux-based. The data is also easily synchronized to another database kept at an external location using the Ethernet or the Wi-Fi connection. The two clouds that have been used are the Microsoft Azure cloud platform and Google Firebase. The mobile platform allows the user to study the conditions of the farm in real-time and handle any condition if there is a critical situation. For future scope, if a drone is attached to the system, then diseases or fungal infection of the crops can be detected.

Devi et al. (2020) proposed a general block diagram of a precision agriculture technique that can be built using Machine learning and IoT. The first block is made up of detectors that can take measurements of the temperature, soil moisture content, and relative humidity. The next block, the second one consists of a disease-detection camera. Image recognition methods are used in this block. Drones and farm robotics are used to inject chemicals and fertilizers in the third block. All of this data is then sent to an Arduino, and then to a Raspberry Pi. In between Raspberry Pi and the node, information is shared. These sensors provide the farmer with crop knowledge on a daily basis. He will be able to take the appropriate step by delivering a response to the appropriate machines and the individual who worked the system. If this is not the case, the machine will take steps to get the crop to ideal circumstances. This model can be used for a Greenhouse or smart irrigation system. For the irrigation method, the proposed ML technique that can be used is the k-means clustering method. Support Vector Machine (SVR) has been proposed to measure the weather conditions during a fixed interval of time.

Suma et al. (2017) proposed a design in which the moisture, PIR sensor, temperature sensors are attached to a micro-controller. An LED and a buzzer are attached to the system. The buzzer gets switched on and the LED starts blinking to make the farmers alert that the threshold values set by the farmers for the sensors have been crossed. When such an alarm is sent to the farmer, the power gets switched off immediately. The entire system is connected through the RS232 connection. The system has been designed in two ways—automatic and manual. In the manual mode, the farmer has to manually turn the power off and on using the android application that has been built; whereas, in the automatic mode, everything operates automatically. The design has been implemented using the GSM module in both modes.

Hartung et al. (2017) proposed that according to research, WSNs have been used in the agricultural sector for some time to improve remote control of agricultural products and services, according to the study. WSNs' potential to increase productivity and minimize duplication has greatly increased. With the increase in the potentiality of WSN, the sensors and deployments are very successfully integrated with IoT. Many more systems are built which provide smart fertilization to treat diseases of crops and pesticide issues. Robotic technology has also advanced in this sector in which bots are built. An agricultural bot that has a multispectral detection system driven by a remote IoT disease control device is also gaining a lot of popularity with specific spray nozzles. They can more accurately identify and cope with pest challenges. As a result, this kind of IoT-based pest control scheme helps to restore the natural environment while still lowering total costs.

We may infer from the above literature review that certain systems have been established that identify environmental conditions such as temperature and soil conditions of crops, and that appropriate measures are taken in those systems to aid crop development. Other devices are in place to track the protection of the areas. Some irrigation systems are designed to keep the system in check. Existing systems often sometimes lack the capability to detect whether the crop has been affected by some disease or some fungus. Some systems lack an ML model which will detect the anomalies of the sensor readings and send an alert to the user. The proposed device uses sensors to keep track of not only the soil contents but also the security of the crop area.

This book chapter's proposed model will more accurately conduct intelligent and efficient irrigation, soil moisture content, and soil pH value classification. Cloud computing can then be used to evaluate data and make predictions about crop quality, crop protection, and crop yield, among other things. It will also be useful to carry out operations and automate tasks when required in the future, such as irrigation as needed and notifying the farmer of any irregularities. The system also has a monitoring component to detect fungal infection or diseases of the crops through a sensor attached to a drone that captures an image. All these collected information and analysed and predicted values are stored in the cloud. The ML will be able to detect any anomalies in environmental parameters in order to accomplish the objective of early detection, and the environmental variables of the next time period will be predicted by observing the agricultural temperature and humidity parameters of the current time value. The LSTM model will be able to forecast the estimated environmental values of the next moment. With the addition of machine learning, the IoT technology can run in parallel and concurrently with current hardware systems. This will result in a crop management solution that is successful.

## 5.3 Proposed Model

The model that has been proposed in this book chapter can guide the farmers about the conditions of the soil, the temperature of the environment, and can send an alert to the farmer to protect the crops from any disease or fungal infection. A phone application is built to which the processed output is sent to the user's phone via the Twilio API.

We present our proposed architecture for building a smart agriculture system using IoT technology and machine learning algorithms in this section. Our model attempts to overcome the most popular harvesting problems. These include appropriate irrigation of the water, proper soil quality control, proper field monitoring, avoiding disease attacks on crops, and increasing harvesting profitability. This model's inspiration is to solve obstacles in a more intelligent manner so that farming becomes smoother for farmers due to technology, and it also becomes less time-consuming with substantially less manpower and resources, and with a large rise in crop growth. The framework of the proposed model is shown in Fig. 5.1.

The proposed model has 'n' number of primary nodes in which the sensors are connected to an Arduino Uno micro-controller. These sensors collect the information regarding the environment and the soil using the following sensors: pH sensor, the LM393 soil moisture sensor, the temperature and humidity DHT11 sensor. The primary node also has an Unmanned Aerial Vehicle (UAV) which helps in monitoring the entire field to detect fungal infection and diseases of the crops by clicking pictures using an RGB-D camera. The working principle of an individual *i*th primary node is shown in Fig. 5.2.

These collected data are sent to the coordinator node which has Raspberry Pi and a water pump through the nRF24L01 transceiver. The Raspberry Pi sends data to the IoT hub using the GSM module; which sends the data to the database and the ML algorithm (LSTM) for prediction. All of these data are stored in the cloud. The analysis of data, predicted data, and all details are all available on a phone

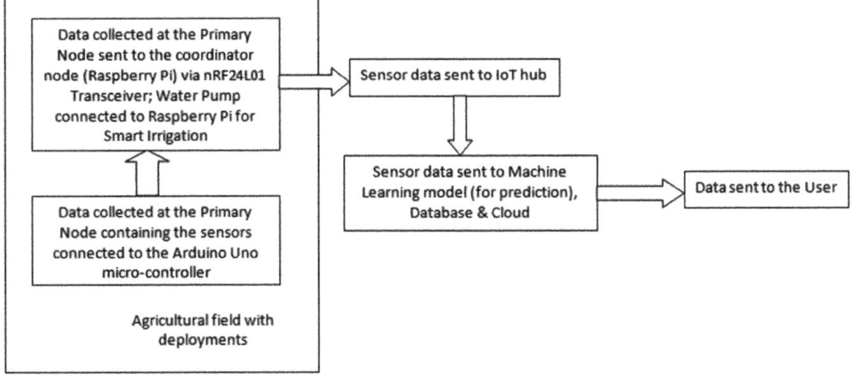


**Fig. 5.1** Framework of the proposed model

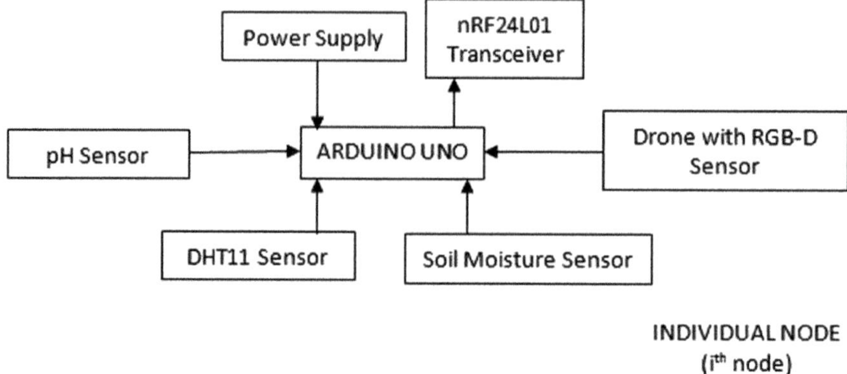


**Fig. 5.2** Working principle of an individual *i*th primary node

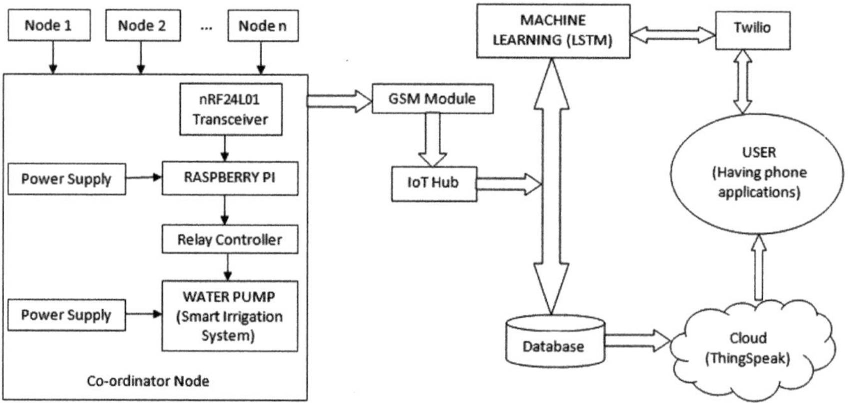


**Fig. 5.3** Working principle of the individual nodes with the coordinator node and the cloud

application in the user's smartphone. The working principle of the individual nodes (primary nodes) with the coordinator node and the Cloud is shown in Fig. 5.3.

### *5.3.1 pH Sensor*

The composition and consistency of the soil are important factors in crop health. As a result, pH management is crucial. Plants thrive better when the pH is between 5.5 and 6.5. Several species of plants need more acidic soil, while others require more alkaline soil. pH management maximizes fertilizer production by regulating the bio-distribution of soil nutrients. The pH of soil influences the occurrence of toxic elements in the soil, the composition of such soils, and the behaviour of soil microorganisms. In greenhouses and nursery growing pots, pH observations of organic substances are necessary. pH should be tested right away to ensure that the specimen purchased has the required pH. The consistency of irrigation water is a crucial consideration. Other inconsistencies can be present if the pH value is significantly lower than pH 7. Any pH sensor would require a proper flow of electricity as it depends on the voltage tester to find out the hydrogen level along with the pH of the component.

The pH-HH-SOIL metre has a multilevel LCD screen that shows pH with a range from -2.00 pH to 16.00 pH along with a resolution of 0.01 pH. The metre shows values with a precision of ±0.02 pH. It can also provide temperature readings at the same time. On the main LCD, the pH value (automatically adjusted for temperature) is shown, while the temperature of the solution is shown on the secondary LCD. Other characteristics of the pH-HH-SOIL metre that are used in more expensive portable devices include automated adjustment, buffer detection, and temperature adjustment. The pH electrode provided is a design made up of glass; an improved electrode of the pH electrode that already exists; with an incorporated temperature detector engineered specifically for calculating the pH of the soils. There are also indicators for the adjustment and reliability state upon on LCD. Since the proposed framework will use a digital pH sensor, so the pH sensor is attached to an Arduino Uno micro-controller for collecting and monitoring the pH information of the soil.

### *5.3.2 Temperature and Humidity Sensor*

Both humidity and air temperature are detected, monitored, and recorded by the humidity sensor. Growing crops necessitates the accurate measurement of humidity and the regulation of heat or cold. Humidity sensors come in three varieties: resistive, capacitive, and thermal.

We used the DHT11 sensor in this developed system which is a low-cost, commonly accepted sensor that detects temperature and humidity information from the surrounding atmosphere and generates a calibrated digital display. It also includes a Negative Temperature Coefficient (NTC) temperature monitoring unit, which is linked to an 8-bit high-performance micro-controller that measures and outputs temperature and humidity values as sequential information. The range of temperature that can be read via this sensor is from 0 to 50 °C and the range of humidity that can be read is 20–90% with a precision of ±1 °C and ±1%. It utilizes a capacitive humidity sensor and a thermistor to measure the humidity level and sends a digital signal to the micro-controller (analogue input pins are not required). Long-term reliability, rapid response, relatively inexpensive, and high anti-interference potential are all advantages that make this sensor favourable for IoT projects. Furthermore, it has an accurate calibration with a low error rate. The sensor is unusual in that it has a resistive sensation for wet materials, making it ideal for use in farming areas; and hence making it very suitable for this proposed model.

### 5.3.3 Soil Moisture Sensor

Soil Moisture sensors are used mainly for irrigation system, gardening, in controlled environments, etc. A soil moisture sensor is employed to provide a sense of how much water is in the soil. In order to irrigate the agricultural sector, a traditional farmer must analyse the soil moisture content. The LM393 comparator circuit is used by this soil moisture sensor along with two probes to determine if the humidity is higher or lower than the ideal threshold humidity (which has been set by the farmer). The sensor's capabilities include a customizable threshold level and a two-state binary result (low or high). When the sensor is connected to the micro-controller and the probe is inserted into the soil, it conducts more energy, which ensures the resistance is small and the moisture level will be high if there is more water content in the soil. This sensor's module consists of resistors, capacitors, power, LED status integrated into a circuit. The requisite current is less than 20 mA, the maximum voltage required is 5 V, and this sensor works in a temperature between 10 and 30 °C. Soil moisture sensors help in controlling water flow, thus saving energy and helping in better yields of crops. Since this sensor is easy to be used micro-controllers and is also affordable, small, and available easily, so this sensor has been used in this proposed model.

### 5.3.4 RGB-D Sensor

To reflect and study the soil, the proposed model uses a low-cost specialist RGB-D sensor. RGB-D sensor is a category of depth sensor that works in conjunction with an RGB (red, green, and blue colour) sensor camera. Many difficult tasks, including such as object recognition, picture sorting, pose estimation of a pose, visual recognition, semantic segmentation, analysis of shapes, image-based rendering, and many others, can benefit from the use of depth details and analysis. The infrared sensor provides depth data which is synced with a configured RGB camera, which generates an RGB picture with a depth correlated with every pixel. They will add detailed depth information (connected to the proximity to the detector) to a traditional picture on a per-pixel base. It could be used with either rotary wings, such as a multi-rotor device, or a fixed-winged drone.

### 5.3.5 Arduino Uno

Arduino Uno is a micro-controller. The micro-controller is based on the ATmega328P (datasheet). Using this micro-controller is advantageous as it is quite cheap and the hardware along with the software is open source. The programming is quite easy and the IDE of the software is compatible with any Operating System. It simplifies the hardware and the software's amount and provides a lot of libraries making the programming quite simpler. Arduino Uno allows both serial communication and programming over USB. The ATmega328P is an 8-bit improved RISC (Reduced Instruction Set) processor that comes in a 28-pin Plastic Dual-In-line Package (PDIP) package. It has 2 KB of SRAM, 6 analogue pins with 10 bits, and 14 input and output optical pins. A 16 MHz frequency resonator, a 5 V power port, and a USB interface are all included on the board. This will handle voltages ranging from 7 to 20 V. It can be simply connected to a computer or an AC-to-DC adapter can be used to get it started.

### 5.3.6 *nRF24L01 Transceiver*

nRF24L01 transceiver has a lot of applications like wireless data communication, surveillance, industrial sensors, alarm and security systems, etc. It is basically a radio transceiver and is a single chip whose transceiver module board consists of Antenna Trace (on PCB), nRF24L01 Transceiver Integrated Circuit (IC), along with a 16 MHz crystal. The module board also has communication and processing pins along with other passive components like the demodulator, modulator, power amplifier, etc. It is a 2.4 GHz wireless transceiver that can send data at up to 2Mbps as well as operates on a 1.9–3.6 V supply voltage. The end-node device consists of the sensors and the Arduino Uno sends sensor information to the Raspberry Pi through the nRF24L01 transceiver. Since it has a collision avoidance feature, it can handle sensor data from several nodes at an identical time.

### 5.3.7 *Raspberry Pi*

Raspberry Pi is a small Single Board Computer (SBC) based on an ARM and is very low in cost. It has both the facilities for Wi-Fi and Bluetooth. The proposed model uses Raspberry Pi 4 Model B and the sensor data reach it through the nRF24L01 transceiver from the Arduino Uno. This model has been used as it is a quad-core model and is comparatively faster and more capable than its predecessors. This module will send the transformed digital equivalents of the received variables to any cloud-based storage area over the internet. Other advantages of this model include online surfing, which allows the user to access the website designed specifically for this device, and the extra memory that is present allows the user to quickly move between smartphone applications and large websites without slowing down or lagging. The core programming languages are Python and Scratch, but it also supports other programming languages making it user-friendly. It is also advantageous as it allows the user to choose the operating system that he can use.

### 5.3.8 *Machine Learning*

Artificial intelligence includes Machine Learning (ML). ML involves rather than programming a machine to solve the problem directly, it learns from the examples given. Machine learning algorithms create a model-dependent, on reference data referred to as the 'training data', in order to make projections or decisions without being directly implemented. Depending on the input information, ML models identify patterns and then use statistical study and analysis to forecast the outcome within a reasonable range. ML algorithms are most often divided into two categories: supervised and unsupervised learning. The individual must have both input and goal features for supervised algorithms. The learned algorithm is then used to infer new data until the training is complete. The individual does not need to have the goal in unsupervised instruction. For our proposed model, we have used the Long Short-Term Memory (LSTM) ML algorithm. LSTM is a deep learning architecture that uses an artificial recurrent neural network (RNN). LSTM has feedback relations, unlike normal feedforward neural networks, which make it advantageous for use. As there may be lags in intervals of uncertain length between significant events in a time series, LSTM neural networks are preferable for categorizing, analysing, and predicting outcomes dependent on time series data.

### 5.3.9 *Thing Speak*

ThingSpeak is an open-source cloud-based service portal that allows users to collect, analyse, visualize, and share data. Dashboards give users a holistic view of their most relevant information in one location, and they are updated in real-time in the cloud and the database. ThingSpeak often easily records the data sent, allowing users to monitor their devices or appliances remotely from anywhere and view the information on either smart computer or smartphone applications or Web browser, making it ideal for IoT ventures. It works with a variety of open-source software, including MATLAB Visualizations and Analysis, ThingTweet, TimeControl, etc., and also allows hardware to link to itself. These features aid in the processing of data that has been submitted. In our proposed model, ThingSpeak acts as a central server, preserving all metadata. The sensor data is collected in a single location in the cloud and stored and sent to the ML algorithm for training and predicting the weather, enabling us to easily find it for online or offline research, making it easy to incorporate it into our developed framework.

#### 5.3.9.1 Twilio

Twilio Customer Engagement Platform is a developer framework for messaging. The programmable application programme interfaces (APIs) from Twilio are a series of building blocks that developers can use to create the exact consumer interactions they want. Using capabilities such as WhatsApp, text messages, audio, video, mails, and now even IoT, the Twilio Customer Engagement Platform can be used to create almost any interactive experience through the consumer experience. The phone application that activates the ML Studio API starts the inference process, where the potential parameters of the soil are evaluated and plotted with the appropriate temperature and other environmental parameters from the literature survey.

## 5.4 Methodology

The proposed system consists of the following main components: (1) the cloud, (2) remote WSN, and (3) the mobile application. The remote WSN consists of an array of sensors that will monitor the environmental conditions for the crops. These arrays of sensors include the pH sensor, the LM393 soil moisture sensor, the temperature and humidity DHT11 sensor, and the RGB-D sensor attached with a drone.

Ideal parameters to grow a completely nourished crop includes the temperature of the soil, air temperature, the level of moisture and humidity present along with other parameters like the soil's texture, soil drainage, conditions of the root, etc. These parameters help in forecasting which crop is suited or which crop requires the necessary favourable conditions to grow.

To measure the favourable parameters for our proposed model, the above sensors are used and connected with a micro-controller. The information that is obtained from these sensors is collected through a WSN and then uploaded to the cloud using a gateway. For this proposed model, other factors like the soil's texture or depth or drainage have not been considered as these factors can be considered as less invariant and also this information can be obtained while doing manual testing of the soil. The details or information that is obtained from these sensors are collected through the WSN and stored in the cloud which is the database.

The model has two types of nodes in which two different micro-controllers are present. They are a primary node and a coordinator node. Each primary node consists of an array of sensors that are connected to the Arduino Uno micro-controller. The temperature and humidity of the atmosphere, as well as the pH and moisture content of the soil, are all recorded by these sensors. This micro-controller is connected with a power supply and nRF24L01 transceiver.

The Arduino Uno also gathers data and sends it to the Raspberry Pi, which is collected by an RGB-D sensor on a drone or UAV (Unmanned Aerial Vehicles). It is done to make this model a monitoring system in which a drone or a UAV fitted with an RGB-D sensor is used to verify the condition of crops so that insect attacks, crop health degradation, veld fires, flooding, or landslides can be effectively noted and appropriate actions taken. The UAV is powered by a battery that provides enough energy for it to travel and move about in the appropriate areas.

Any utilization of urea nitrogenous-based or ammonium-based fertilizers could indeed cause soil to become acidic, necessitating daily soil supervision. Ranges for assessing water quality include: pH range of 6–8.5 is acceptable and it can be used without difficulty;—pH of 5–6 or 8.5–9 is adequate but sensitive crops can have issues;—4–5 pH or 9–10 pH is considered to be in short supply and should be used with caution and wetting the plants should be avoided;—pH4 or pH > 10 is considered to be very rare and so other variations must be detected by chemical examination. The value of the pH of the nutrient solution (which consists of water and fertilizer) must be appropriate for the crops. Soil pH values (observed in a mixture of water and soil) indicate that it is highly acidic below 5.0, mildly acidic between 5.0 and 6.0, neutral between 6.5 and 7.5, and heavily alkaline between 8.5 and beyond. The optimal pH range for promoting plant growth and the microbial life associated with it is 5.5–7.0. The availability of vital nutrients to plants and the behaviours of advantageous bacteria and microbes may be affected by soil pH if outside of the neutral zone. Hence, maintaining the pH of the soil and water is very important for cultivating crops. To display the readings of the sensor on the panel, digital pH sensors need an analogue-to-digital converter. On the other hand, in analogue pH sensors, the pulse received activates a coil of electromagnetic, which causes the needle to spin, resulting in a pH measurement on the metre.

Hence, Arduino Uno serves as the micro-controller in each and every node. It collects data from sensors and sends it to the coordinator node which is the Raspberry Pi. The end-node device which consists of the sensors and the Arduino Uno sends sensor information to the Raspberry Pi, which requires a voltage of 5 V, through the nRF24L01 transceiver. To link Arduino and Raspberry Pi, we have used a pair of nRF24L01 transceivers in the proposed model. This Pi acts as the coordinator node and also uploads the data onto the cloud from where the data is used for ML. The LSTM architecture is built on a recurrent neural network (RNN) architecture that recollects data at random periods. This method is well adapted for time series data set classification and estimation, and hence has been used to predict weather conditions like temperature, etc.

The RGB-D camera takes and transforms photographs in real-time. Instead of using hyperspectral imaging, the UAV is using a traditional multispectral technique to save money. The benefit of this method is that farmers can easily use UAVs with multispectral imagery during the cropping season. It is possible for the farmers to predict agricultural productivity owing to multispectral imaging capabilities. The UAV's trajectory is modelled using the field's local map, which is maintained on-board.

The primary individual nodes use the Zigbee wireless connection in their router. Since Zigbee is cost-effective and is low in power consumption, it gives the advantage to use in IoT projects and hence has been used in this model. Though Zigbee supports both tree and star networking, Mesh topology has been used in this model as then the routers can be extended for communication at the network level. To read data from sensors and for its transmission, the defined rate of 250kbit/sec of Zigbee is used.

The data from the primary nodes are collected and sent to the coordinator node, the Raspberry Pi which then sends the sensor readings to the IoT. The Pi is connected to another nRF24L01 transceiver which allows it to receive the sensor readings from the primary nodes. The Pi is connected to a water pump and a power supply. Raspberry

Pi plays a crucial role in the concatenation of the sensor readings data by uploading the entire bulk of sensor information to the IoT via a gateway. The gateway that has been used in this framework is the GSM module.

As the data in bulk reach the IoT hub, the user can send continuous queries to the system for analysing and performing actions on real-time data. As the data stored in the database is the time series data, so the data set is sent to the ML studio to train a model. The trained model becomes more accurate and precise by training on more data that is acquired with time.

To achieve the goal of early warning, the LSTM model is able to detect any anomalies that are present in environmental parameters; and the environmental parameters of the next day or the next week are forecasted by studying the agricultural climate parameters of the current time and the present day. For example, if the temperature sensor reads a wrong temperature due to any issue or the pH sensor reads a wrong pH value or the soil moisture sensor has an erroneous reading, then the LSTM model is able to predict and forecast the approving temperature or pH value or the required soil moisture content. With the addition of machine learning, the IoT technology in our model can run in parallel and concurrently with current hardware systems. All these data are stored in the cloud and are sent to the smart device used by the farmer so that he can monitor from a remote place. The ThingSpeak is used for storing, aggregating, analysing, visualizing, sharing, and retrieving data from the Raspberry Pi board. The dashboard of ThingSpeak is beneficial to the user as it provides an overall view in a single place to the users in real-time and the important details are also updated in real-time.

Along with this, every primary node records the soil moisture content. There is a pump attached with the coordinator node which gets switched ON after getting triggered if the soil moisture content is recorded as less than the threshold range set by the user.

A mobile application is designed which show the sensor readings stored in the cloud for a particular day, shows images that are captured using the RGB-D sensor; also displays the future parameters of the soil after training from the dataset and shows if the smart irrigation system status, that is, if the water pump is ON to maintain the soil moisture level to reach its optimum value. All these processed outputs are sent to the user's mobile via the Twilio API. This system can assist consumers in increasing production capacity while also reducing the risks of crop loss. The flowchart of the working of the primary (individual $i^{th}$) node is shown in Fig. 5.4. The Flowchart of the working of the Coordinator node and the entire proposed model is shown in Fig. 5.5.

## 5.5 Performance Analysis

This proposed model is an ML-based smart farming model that helps farmers in making better productivity in the field. The LSTM model was used to forecast the atmosphere's temperature and humidity for the coming time period. The sensors used

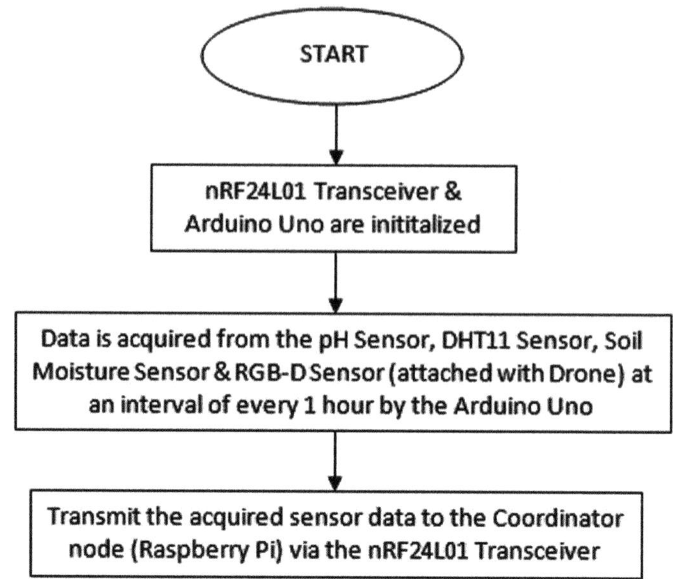


**Fig. 5.4** Flowchart of the working of the primary (individual *i*th) node

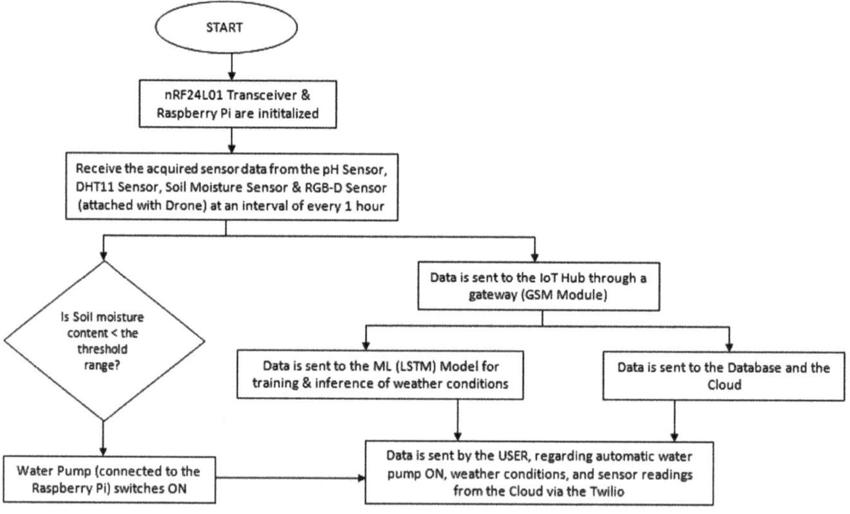


**Fig. 5.5** Flowchart of the working of the coordinator node and the entire proposed model

are accurate and also affordable. The model improves the productivity of harvesting by almost 20%. The use of a smart irrigation scheme, in which the pump is turned on only when the soil moisture content is less than the threshold value set by the farmer, reduces water consumption by nearly 20%. The cost of labour of the harvesting field also gets reduced to almost 55–60%.

To evaluate the model, we have collected the time series data for different time periods for the temperature and humidity DHT11 sensor for sensing the two different environmental parameters in different environmental conditions. The sensors were kept stable and were not in locomotive. For our sample data set, we have collected 160 sample values collected from 10 DHT11 sensors attached to 10 Arduino Unos, which has been passed to 1 Raspberry Pi node, the coordinator node (that is, 10 primary nodes have been used to get the sample data). We remove the missing values and keep the sensor data separate for temperature and humidity, not inter-mixing them.

The prediction of temperature data from LSTM model with sensor data is shown in Fig. 5.6. We have the data from the temperature sensor, where the sensed temperature is in Celsius (°C). The blue line indicates the data read from the sensor and the orange line indicates the data predicted by the model, based solely on the knowledge of the sensor data. The data has very steep curves indicating a sharp change in temperature. The model which had only seen nearly the same data for a long time takes some time to adapt to these changes. It adapts well during the lower trends till the data settles down at a certain temperature, where it misses the original temperature but soon latches on to the original data trend.

The prediction of humidity data from LSTM model with sensor data is shown in Fig. 5.7. We have a sequence of data taken from the humidity sensor data. This portion of data highlights one of the main aspects of our model—approximating. When the data is too noisy, the model tries to keep the data output in a stable feed,

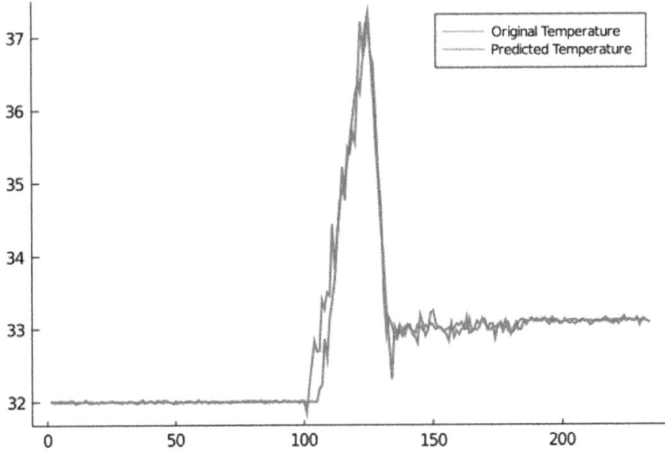


**Fig. 5.6** Prediction of temperature data from LSTM model with sensor data

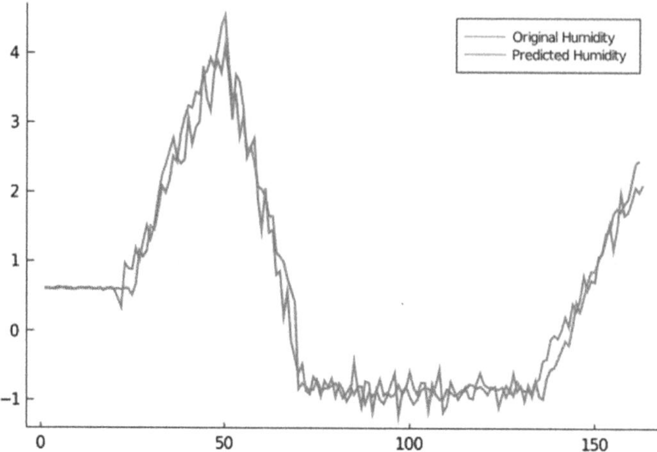


**Fig. 5.7** Prediction of humidity data from LSTM model with sensor data

lowering the amount of impact the noise has. These approximations tend to become smoother and smoother with time, as the model sees more and more data.

These are some experimental results where we compare the data from the sensor and the data predicted by our model at each time step. We have chosen to specify the more interesting portions of the graph that involves more crests and troughs. On the other hand, it is also meant to add qualitative insight into the accuracy of the live model.

With the LSTM model predicting with an accuracy of almost 90–92%, our model on average performs 15–20% better compared to other existing models. Farmers can make smarter harvesting decisions in the future by using predictive analysis of the observations collected by the sensors. This framework's sensors are widely available and affordable, allowing them to be deployed anywhere in the country. The sensors chosen have accurate measurements that can be used in a number of agricultural soil simulation applications. With Zigbee technology being used in a Mesh topology, the connections also cost less and the system consumes low power. We have used ThingSpeak as the cloud platform, as it is very easy to be integrated with IoT devices and has many advantages like displaying and analysing data using MATLAB.

## 5.6 Future Research Direction

As future work, this framework can be further designed to incorporate blockchain technology. Our current model is confined to the collection of sensor data, analysing them, and prediction using the LSTM model. The model can be developed to make all transactions starting from the harvest to the supply market. Making this model grow to large-scale farming can help farmers benefit and save resources.

Licensed low-power wireless access (LPWA) networking protocol is very cost-friendly, and hence can be used in further developing the model as it will cover plenty of geographical locations. For our future work, we can incorporate fog computing, in which the sensors initially communicate to the devices residing at the farmer's place. The analysis reports prior to harvesting will be stored in these computer's databases. The analyses of the data through the ML algorithms are to make using these data collected. Hence, using fog computing not only the intervention of the cloud will get reduced, but the system will be much lower in cost along with being a faster, more secured, and much more reliable system.

Our model currently just makes a prediction of the sensor readings for the upcoming week. This model can be further developed by making the ML algorithm to predict the readings of the sensor for the upcoming year, according to which the model will suggest which crop to be harvested in the next year. Along with this, a comparative study can be made in this model by incorporating different ML algorithms instead of just LSTM. The model can be further utilized to analyse the information generated by the LSTM model to include a drainage system or pesticide spraying agribots, which will also help in better protection from environmental disasters and give a better agricultural processes' automation.

Exposure to a lot of high temperatures can make the batteries of the sensors worn out. Rechargeable batteries that are attached to solar panels, which will give us the desired power for the sensors instead of a continuous power supply, can be a further development of this model. Making this system a cost-effective model based on solar-powered rechargeable batteries can be a future scope. The Arduino micro-controller board was used in this model. Good networking and applications also help to get better performance by increasing processing speed. As a result, a safer alternative approach is to use specially built MCU boards to optimize the system's usage, which would also minimize configuration costs and power consumption significantly.

The data communication from one end to the other and the sensors are usually susceptible to getting distorted due to lateral channel attacks, which can lead to false information and affect the agriculture process. Using lightweight cryptographic algorithms to secure the data transmission and the readings of the sensor can be further future scope for this model. Middleware architecture in the future will further guarantee a framework with end-to-end security. A process to authenticate users in the system can be further incorporated to avoid securities and middleware-leakage; hence will help to maintain the confidentiality of the process.

## 5.7 Conclusion

A farmer has to perform a variety of activities in a harvesting area, including irrigation, soil health monitoring, productivity control, field protection, keep track of the environmental factors, and so on. The farmer is served by this proposed model 24 × 7. Crop growth patterns and ecological criteria provide technical recommendations and optimum agricultural prevention steps for proper yield. As a result, continuous, large-scale monitoring of crop growth in response to changing environment is needed. The model will track the field even when the farmer is not present, so the farmer does not need to be present all the time. Our model will intelligently manage various harvesting problems due to the IoT and ML. These include sufficient irrigation, proper soil quality and field monitoring, avoiding diseases on crops, and increasing harvesting production. As a result, the system guarantees a significant amount of time and resources to be saved. Hence, the model allows the farmer to save a lot of resources and focus on increasing the productivity of the field. This model is cost-effective as the sensors and the connections used are at an affordable price. Also, the cloud used is open-source and performs analysis and visualization of the data making it an advantage. The attack of fungus, pests, or any other disease attack can be handled well using the RGB-D sensor attached to the UAV, which will allow the farmer to keep in check which fertilizers and pesticides are to be used. The health of the soil and the crop can be monitored well using the other three sensors—the pH sensor, the LM393 soil moisture sensor, the temperature and humidity DHT11 sensor; which measure the pH of the soil, the moisture content present in the soil, and the temperature along with the humidity of the surrounding, respectively. The soil moisture sensor is also highly efficient as it helps to maintain a proper irrigation system, thus saving water. The proposed system is a learning-based approach to classification. The prediction becomes more accurate with time and helps the farmer in realizing the entire soil's parameters and its pattern for the future. Thus, the farmer becomes alert if the soil parameters change and tries to maintain the parameters as mentioned by the LSTM model. In terms of precision planting, this system performs better than the current systems. When compared to traditional systems, this method is more reliable and produces improved performance. The emphasis of this method has been on executing tasks in a timely, cost-effective, and reliable manner. As several broadband carriers extend fair Internet access in rural areas, this system has become very cost-effective and stable. The advantages of using the cloud network are that data can be viewed from virtually anywhere and at any time. With the implementation of sending updated details to the smart phones' application in our model, along with sending the back-end data's analysis for prediction of weather and soil conditions, this framework gives a better crop yield and hence, in short, is very robust.

## References

Ayaz M, Ammad-Uddin M, Sharif Z, Mansour A, Aggoune EHM (2019) Internet-of-things (IoT)-based smart agriculture: toward making the fields talk. IEEE Access 7:129551–129583

Devi YSS, Prasad TKD, Saladi K, Nandan D (2020) Analysis of precision agriculture technique by using machine learning and IoT. In: Soft computing: theories and applications. Springer, Singapore, pp 859–867

Garg H, Dave M (2019) Securing IoT devices and securelyconnecting the dots using rest API and middleware. In: 2019 4th international conference on internet of things: smart innovation and usages (IoT-SIU). IEEE, pp 1–6

Hartung R, Kulau U, Gernert B, Rottmann S, Wolf L (2017) On the experiences with testbeds and applications in precision farming. In: Proceedings of the first ACM international workshop on the engineering of reliable, robust, and secure embedded wireless sensing systems, pp 54–61
Jyothi PMS, Nandan D (2020) Utilization of the internet of things in agriculture: possibilities and challenges. In: Soft computing: theories and applications. Springer, Singapore, pp 837–848
Joshi J, Polepally S, Kumar P, Samineni R, Rahul SR, Sumedh K, …, Rajapriya V (2017) Machine learning based cloud integrated farming. In: Proceedings of the 2017 international conference on machine learning and soft computing, pp 1–6
Mat I, Kassim MRM, Harun AN, Yusoff IM (2016) IoT in precision agriculture applications using wireless moisture sensor network. In: 2016 IEEE conference on open systems (ICOS). IEEE, pp 24–29
Mekonnen Y, Namuduri S, Burton L, Sarwat A, Bhansali S (2019) Machine learning techniques in wireless sensor network based precision agriculture. J Electrochem Soc 167(3):037522
Saha HN, Roy R, Chakraborty M, Sarkar C (2021a) Development of IoT-based smart security and monitoring devices for agriculture. In: Agricultural informatics: automation using the IoT and machine learning, pp 147–169
Saha HN, Roy R, Chakraborty M, Sarkar C (2021b) IoT-enabled agricultural system application, challenges and security issues. In: Agricultural informatics: automation using the iot and machine learning, pp 223–247
Suma N, Samson SR, Saranya S, Shanmugapriya G, Subhashri R (2017) IOT based smart agriculture monitoring system. Int J Recent Innov Trends Comput Commun 5(2):177–181
Varman SAM, Baskaran AR, Aravindh S, Prabhu E (2017) Deep learning and IoT for smart agriculture using WSN. In: 2017 IEEE international conference on computational intelligence and computing research (ICCIC). IEEE, pp 1–6

# Chapter 6
# IoT Doordarshi: Smart Weather Monitoring System Using Sense Hat for Improving the Quality of Crops

**Harshita Jain, Kirti Panwar Bhati, Nupoor Katre, and Prashant Meshram**

**Abstract** The main aim of the proposed system is to build a system that will sense the environmental parameters in an area of interest for crop quality monitoring. System monitors the environmental parameters in real time and sends the data to a cloud that can be further used for analysis. This system describes a complete infrastructure for environmental monitoring and controlling. The proposed system makes use of low-cost Raspberry Pi and Sense Hat Sensor for monitoring the environment. As the title suggests Doordarshi, the one who is visionary and performs the task so to get the favorable result in future, likewise the proposed system monitors the environmental parameter to help us to take corrective actions for improving the crop quality in the future. The proposed system senses the weather parameters and monitors it in the real time and then email the information (using SMTP protocol) and phone notification (using Pushover IFTTT service) that can be further used for analysis. Cloud service, ThingSpeak, is used for storing and plotting the data. Weather detection system using IoT has been experimentally established to work satisfactorily by connecting the different modules of IoT into a single platform. The designed system not only monitors the real-time data for crop quality monitoring but also sends it into the mail and notification for further analysis. This will eventually help to determine the different environmental conditions and accordingly corrective measures can be taken for improving crop quality.

**Keywords** Crop Quality · IoT · Sense Hat · RPi · Doordarshi

## 6.1 Introduction

Doordarshi is a Hindi word the meaning of which "showing the ability to think about or plan the future with great imagination and intelligence" according to the Oxford Dictionary. Here title suggests Doordarshi which is a visionary IoT System for getting good quality crops by making use of low-cost tools and technologies. The

---

H. Jain · K. P. Bhati (✉) · N. Katre · P. Meshram
School of Electronics, Devi Ahilya University, Indore, India

 
A. Choudhury et al. (eds.), *Smart Agriculture Automation using Advanced Technologies*, Transactions on Computer Systems and Networks,
https://doi.org/10.1007/978-981-16-6124-2_6

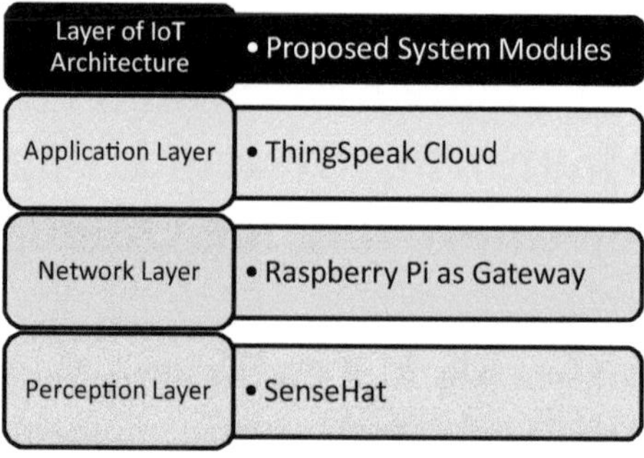


**Fig. 6.1** Proposed system components for the implementation of three-layer IoT architecture

proposed system makes use of techniques where manual intervention is not required for crop environment monitoring.

IoT has found its place in each and every domain of human life. The application of IoT ranges from high-end technological domain to the area which deals with the stakeholders, which are not exposed to technologies at all. Agriculture is one of those fields that has end-users who are not very much exposed to the technology and carry out the process of agriculture the way it was before decades. India is a leading country in terms of agriculture output, it is the demand of the day for the introduction of technology in agriculture also.

The proposed system is making use of different hardware and software components for the implementation of IoT system, which performs the environment quality monitoring in a crop field and based on that, the operation can lead to quality crop productions. Figure 6.1 shows the proposed system components that are implemented at different layers or implementation of Thre-Layer IoT Architecture.

As suggested by Raj K. Goel et al., Smart Agriculture term is an umbrella that covers science, innovation and space technologies. They have talked about the various technologies, which have a positive impact on agriculture sector in developing nations (Goel et al. 2021).

Doshi et al. (2019) present an IoT solution for Smart Farming for enhancing the productivity in farming and have given the solution for remote monitoring of the field using IoT technologies.

Ratnaparkhi et al. (2020) present the importance of the sensor in Smart Agriculture. Sensors are the important parts of an IoT System, and selection of a suitable sensor plays a very important part in developing an IoT System.

Singh et al. (2021) present the scope of smart farming in India. Temperature has a great impact on crop quality. In this paper (Barlow et al. 2015), wheat crop is considered.

Hatfield and Prueger (2015) have discussed that Beyond a certain point, higher air temperatures adversely affect plant growth, pollination and reproductive processes so the maintenance of environment temperature is really required.

Temperature, humidity and other environmental parameters play an important role in crop yield, and monitoring the parameter helps to increase the crop quality and yield (Sawan 2018).

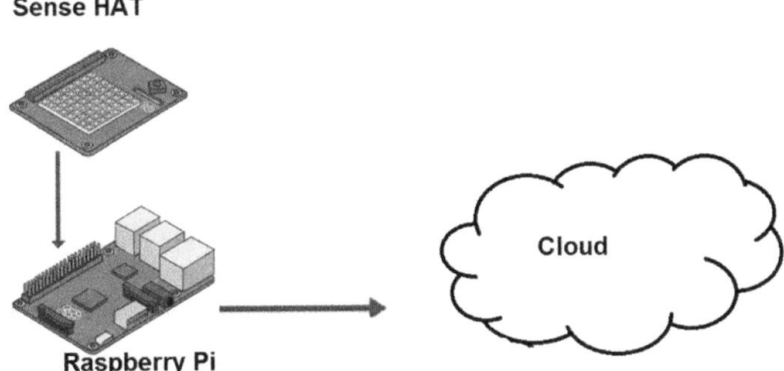


**Fig. 6.2** System architecture

The proposed system makes use of Raspberry Pi as a gateway, which is interfaced with Sense Hat sensor, working as a sensor for monitoring the temperature, humidity and pressure of the environment. The data collected from the Sense Hat are logged on the Raspberry Pi and sent to the cloud for analysis. Here ThingSpeak cloud is used for analysis and visualization of data sent from the Raspberry Pi. Figure 6.2 shows the system architecture.

The parameters that are monitored for the environment are

- Temperature
- Humidity
- Pressure.

## 6.2 Methodology

Raspberry Pi used a Linux-based operating System Raspbian. After installing Raspbianon R-Pi, Sense HAT module has been interfaced with GPIO pin of R-Pi. LPS25H Pressure/Temperature sensors at $0 \times 5C$ address and HTS221 Humidity/Temperature sensor at $0 \times 5f$ address connected through i2c protocol. Installing the sense-hat package will allow the Python module to access Raspberry Pi Sense HAT.

"Sense-hat" is the official support library for Sense-HAT for providing access to onboard sensors and LED matrix. Using onboard sensors measure the value of environmental parameters such as temperature, humidity and pressure. These sensors are connected with I2C protocol on Sense HAT board and have access to GPIO pins of Raspberry Pi. Using Python coding and Sense-hat library value of environmental parameters has been displayed on terminal. Later, this parameter has been displayed on Sense HAT $8 \times 8$ on slide show format.

After monitoring of environment parameters in real time sends these data to Email and push notification as alert/inform messages about the environmental condition surround by field, necessary to detect for good crop quality. SMTP protocol is a standard protocol on a TCP/IP network that provides the ability to send and receive the email. It is an application layer protocol that provides intermediary network services between provider and Server. Using this SMTP protocol, real-time data have been sent as email notification. For push notification, Pushover Service has been used, it is a platform for sending and receiving push notification over phone. On server side an HTTP API is provided for message delivery to the device address.

Along with sending real-time data on Email and push notification, data have been stored on cloud services for further use. ThingSpeak cloud has been used for storing and plotting the data. There is a channel having two fields for receiving the data on cloud. On ThingSpeak MATLAB environment has been used for analysis and plotting the data on ThingSpeak. ThingSpeak also provides the MATLAB visualization for plotting data on 2D plot.

### *6.2.1 Components Used for Implementation of System*

The components used for the implementation of the proposed system can be divided into hardware and software components as given in Fig. 6.3.

**Raspberry Pi**

The Raspberry Pi 3 Model B+ as shown in Fig. 6.4 is used for the implementation of system proposed here. Specifications of Raspberry Pi 3 B+ models are given below (https://www.raspberrypi.org/products/raspberry-pi-3-model-b-plus/).

- BCM2837B0 64 Bit ARM Processor running at 1.4 GHz
- 1 GB SDRAM
- Wireless LAN Support
- Bluetooth BLE Support
- 4 No. 2.0 USB
- Ethernet connectivity

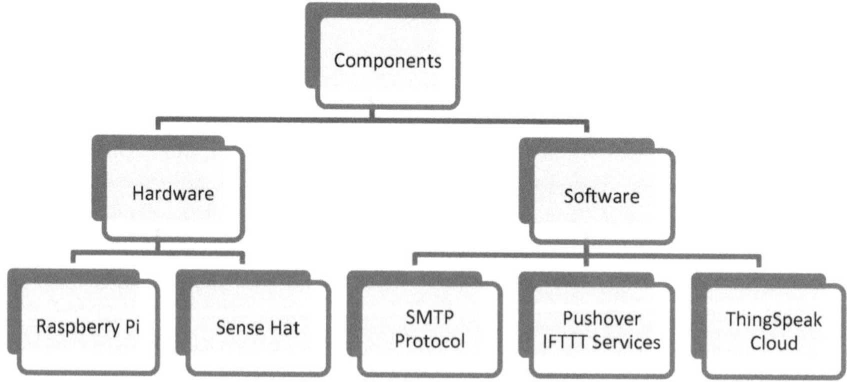


**Fig. 6.3** Hardware and software components required for implementation

**Fig. 6.4** Raspberry Pi 3 B+ model

- 40 Pin GPIO Header
- HDMI Connectivity
- CSI Camera Port
- DSI Display Port
- 4-pole stereo output and composite video port
- Micro SD Port.

### SenseHat

Sense Hat as shown in Fig. 6.5 is used for sending the environmental parameter. Specifications are given below (https://www.raspberrypi.org/products/sense-hat/).

- The sense Hat is a board that can be mounted on Raspberry Pi.

**Fig. 6.5** Sense HAT

- An 8 × 8 LED Matrix, joystick and has following sensors
  - Gyroscope
  - Accelerometer
  - Magnetometer
  - Temperature
  - Barometric pressure
  - Humidity.

**SMTP Client**

SMTP Client is deployed on the Raspberry Pi for sending the email to the mobile phone. It is integrated with email client and having four components.

- Mail User Agent (MUA)
- Mail Submission Agent (MSA)
- Mail Transfer Agent (MTA)
- Mail Delivery Agent (MDA).

**Pushover Service**

To receive the notification of your device, pushover service is used. On the server side, push notifications are sent to the authorized user and client side those push notifications are received and shown to the user and also stored so can be seen even when a device is offline.

**ThingSpeak Cloud**

Data send by device visuals instantly to ThingSpeak, online analysis is done in ThingSpeak with MATLAB code. ThingSpeak uses REST API approach to communicate between IoT device and cloud. The key element of this platform is the channel, which stores and retrieves data generated from sensors via REST API.

Figure 6.6 shows the connectivity between the smart device and Cloud.

## 6.3 System Implementation

Sense HAT senses the parameters (temperature, humidity and pressure) and sends it to Raspberry Pi via GPIO pins.

R-pi processes the data and implements it on the terminal as shown in Fig. 6.7.

Displaying of parameters (temperature, humidity and pressure) on Sense HAT 8 × 8 matrix display done by Python coding on R-pi as shown in Fig. 6.8a–e.

Display of parameter is on 8 × 8 matrix on slide show moving text format.

After reading parameters and analyzing the data on Python editor whether suitable for crop environmental conditions, we'll send the required parameters to email through SMTP protocol as given in Fig. 6.9.

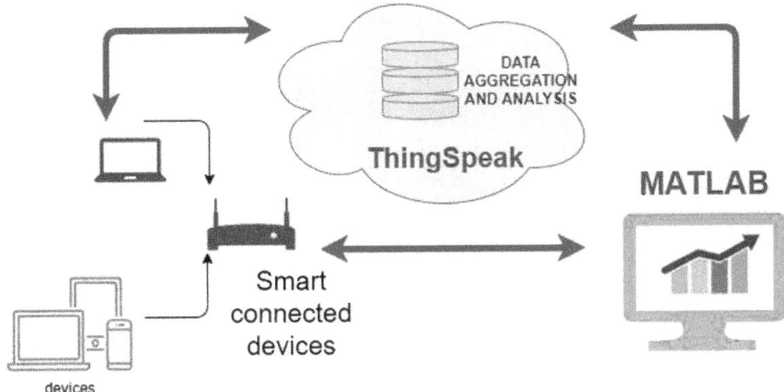


**Fig. 6.6** Connectivity to ThingSpeak cloud

```
pi@raspberrypi:~ $ python iotproject.py
 T=33C, H=59PPM, P=0Pa
 T=33C, H=57PPM, P=950Pa
 T=33C, H=57PPM, P=950Pa
 T=34C, H=56PPM, P=950Pa
 T=34C, H=55PPM, P=950Pa
 T=34C, H=54PPM, P=950Pa
 T=35C, H=54PPM, P=950Pa
```

**Fig. 6.7** Data display on RPi terminal

IFTTT service (if this then that) "Pushover" is used for push notification on phone as given in Fig. 6.10.

Push notification and email alert for showing real-time data and alert message. But we also need to store the parameters and analyze them for further analysis.

For this, ThinkSpeak Cloud is used. On ThinkSpeak cloud using MATLAB Coding analysis of parameters and 2D plotting has been done as given in Fig. 6.11.

Temperature and humdity plotting on a 2D plot after analysis is done, as shown in Fig. 6.12.

## 6.4 Conclusion

IoT Doordarshi has been experimentally established to work satisfactorily by connecting the different modules of IoT into a single platform. This system is able to sense the environmental parameters. These parameters can be sent to user mobile

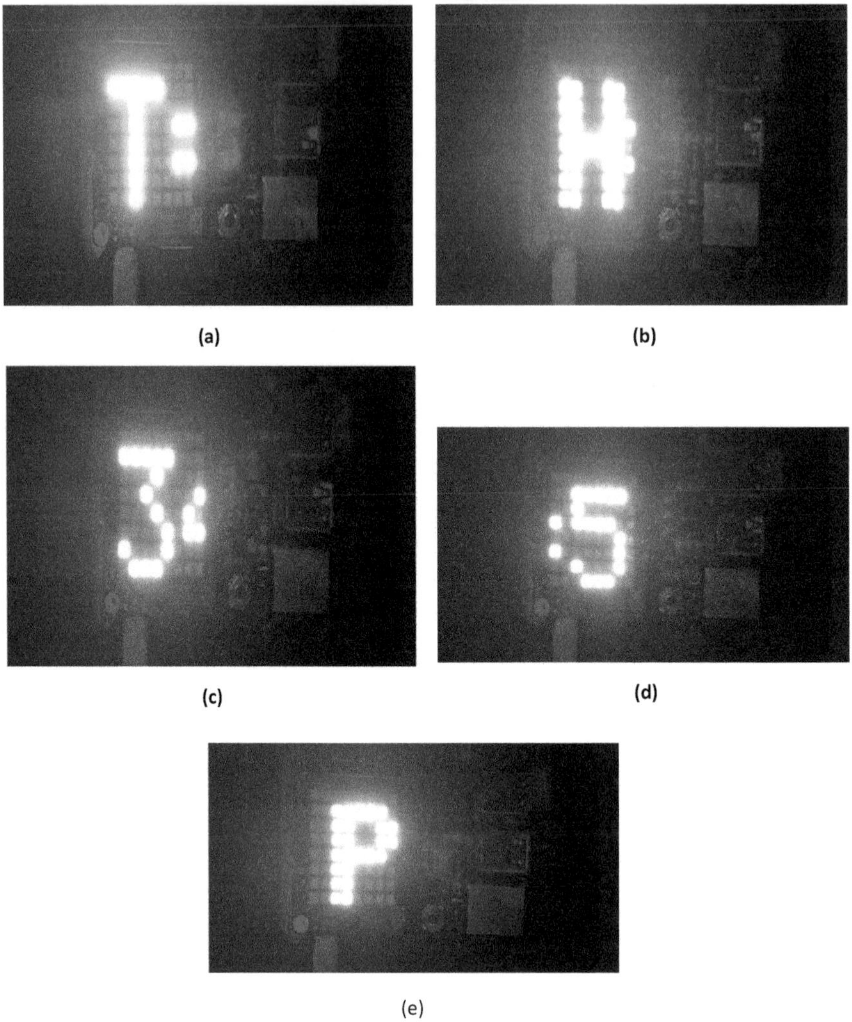

(a)

(b)

(c)

(d)

(e)

**Fig. 6.8** Display of parameter is on 8 × 8 matrix on slide show moving text format

**Fig. 6.9** Sending email through SMTP

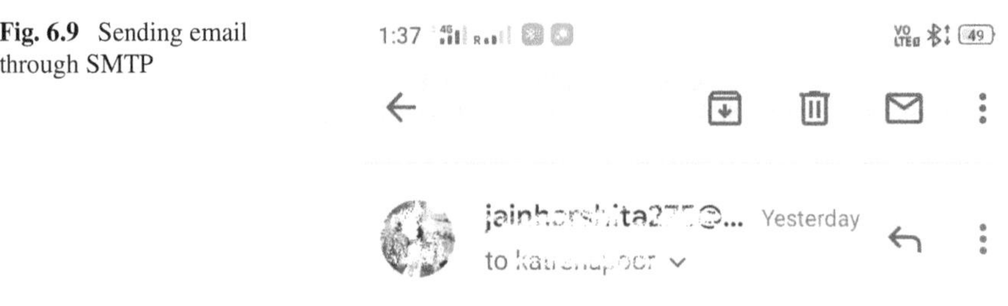

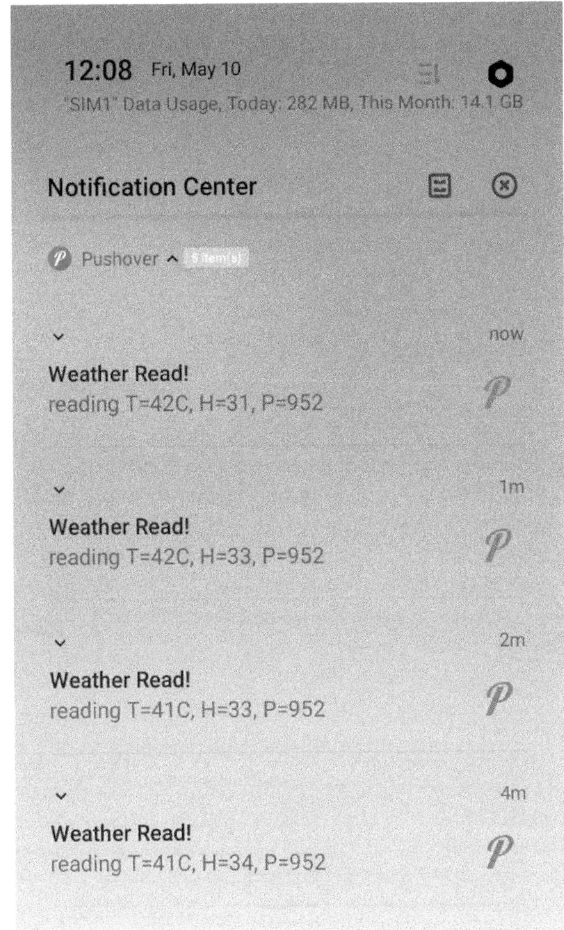


**Fig. 6.10** Push Notification on Phone

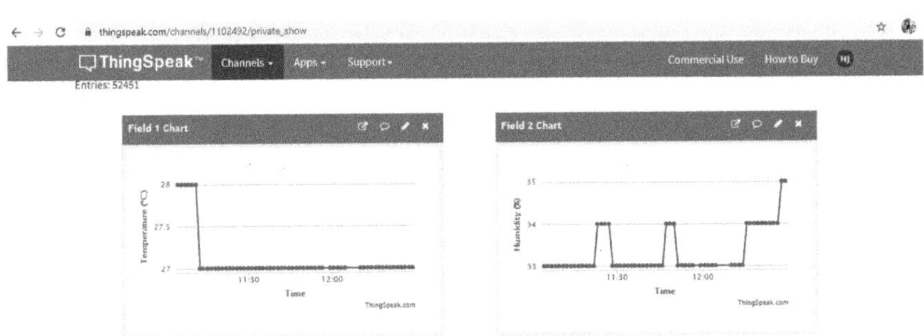


**Fig. 6.11** Analysis of real-time temperature and humidity data

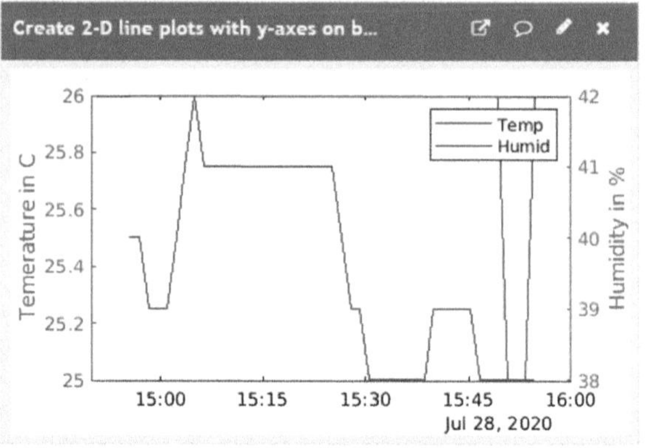


**Fig. 6.12** 2D plotting of real-time data

phone via push notification and email and can be further uploaded on the Cloud for analysis. After analyzing the real-time data on Cloud through predictive analysis a suitable action can be taken for improving the quality of crop.

**Acknowledgment** We acknowledge School of Electronics, Devi Ahilya University, Indore for providing the financial assistance for implementation of the proposed project.

## References

Barlow KM, Christy BP, O'Leary GJ, Riffkin PA, Nuttall JG (2015) Simulating the impact of extreme heat and frost events on wheat crop production: a review. Field Crops Res 171

Doshi J, Patel T, Bharti SK (2019) Smart farming using IoT, a solution for optimally monitoring farming conditions. Procedia Comput Sci: 746–751

Goel RK, Yadav CS, Vishnoi S, Rastogi R (2021) Smart agriculture—urgent need of the day in developing countries. Sustain Comput Inf Syst 30

Hatfield JL, Prueger JH (2015) Temperature extremes: effect on plant growth and development. Weather Clim Extremes 10:4–10

IoT in smart farming .https://www.iotforall.com/iot-applications-in-agriculture/amp/. Accessed 21 April 2020

Raspberry Pi 3 Model B+. https://www.raspberrypi.org/products/raspberry-pi-3-model-b-plus/. Accessed 19 April 2020

Raspberry Pi Documentation. https://www.raspberrypi.org/documentation/. Accessed 20 April 2020

Ratnaparkhi S, Khan S, Arya C, Khapre S, Singh P, Diwakar M, Shankar A (2020) Smart agriculture sensors in IOT: a review. Mater Today Proc

Sawan ZM (2018) Climatic variables: evaporation, sunshine, relative humidity, soil and air temperature and its adverse effects on cotton production. Inf Proc Agric 5(1):134–148

Sense HAT. https://www.raspberrypi.org/products/sense-hat/. Accessed 21 April 2020

Singh PK, Naresh RK, Kumar L, Chandra MS, Kumar A (2021) Role of IoT technology in agriculture for reshaping the future of farming in India: a review. Int J Curr Microbiol App Sci 10(02):439–451

# Chapter 7
# IoT-Enabled Smart Farming: Challenges and Opportunities

**Supriya Jaiswal and Gopal Rawat**

**Abstract** Internet of Things (IoT)-based technologies, cloud computing, big data analysis and computer vision have redefined every sector including smart agriculture. It helped to achieve low cost, high efficiency and high precision farming by using wireless sensor network and communication interfaces. The increase in population has created a huge burden on food industry and limited portion of earth surface is cultivable due to various limitations such as temperature, climate, soil quality, irrigation requirements and topography. In order to resolve this issue, expert intelligent techniques, robots and artificial intelligence algorithms are integrated with IoT to form part of agricultural automation management. Smart farming goes beyond agricultural management task and streamlines data, monitoring and decision-making based on real-time events to introduce new business models related to food industry. Big data are being harnessed to provide predictive insights in farming operations. IoT helps to automate agriculture growth in several ways such as soil mapping and testing, monitoring health and growth of crops, prevention and control of crop diseases, prediction of crop harvesting period, automated irrigation and fertilization, classification and inspection of agricultural products and monitoring farm using Unmanned Aerial Vehicles (UAVs) equipped with image sensor which provide details of crop conditions. However, there are technological challenges needed to be resolved for implementing IoT in new areas. Demand of professionals to develop and utilize the automated agricultural techniques need to be fulfilled. Also, traditional methods of agriculture must be complemented with sensing and driving technologies inspite of completely modifying the complete farming structure. Inclusion of these technological aspects will promote high quality yield of agricultural products and reduce the labor cost and time expenditure.

**Keywords** Smart farming · IoT · Big data · UAV · GPS · WSN

---

S. Jaiswal (✉)
Department of Electrical Engineering, National Institute of Technology, Hamirpur, HP, India

G. Rawat
Department of Electronics and Communication Engineering, National Institute of Technology, Hamirpur, HP, India
e-mail: gopal.rawat@nith.ac.in



A. Choudhury et al. (eds.), *Smart Agriculture Automation using Advanced Technologies*, Transactions on Computer Systems and Networks,
https://doi.org/10.1007/978-981-16-6124-2_7

## 7.1 Introduction to IoT-Based Smart Farming

As per Indian economic survey 2016, agriculture indulges more than 54.6% of the Indian work force and contributed 13.9% to country's GDP. However, this work force reduced to 41.49% in 2020 with increased percentage requirement of crop production. Figure 7.1 shows the report on percentage contribution of various sections in GDP of India for different periods of time (Wagh and Dongre 2016). The limited cultivable land and agricultural resources are the challenging issues which can only be resolved by including smart and automated technology into farming. Internet of Things (IoT) is a crucial technology that has contributed in recent growth of digital market by establishing Machine-to-Machine (M2M) communication. IoT has been versed in different ways depending on its implementation in several application fields such as smart farming, industry, supply chain, retail, healthcare, constructions, energy and transport as shown in Fig. 7.2. Since the global percentage contribution of smart farming-based IoT projects is only 4%, this sector can be more deeply explored to benefit nations' GDP.

One of the definition states "IoT is amalgamation of people, technology and devices with sensors network. The overall integration of IoT with human beings is to achieve communication, technological enhancement and communication to make real-time decisions (Srbinovska et al. 2015). It has found the application in several domains, one of which is agro-industrial sector. The major research in the context of IoT-based smart farming is focused on application domain such as monitoring (62%), control (25%), logistics (7%) and prediction (6%) (Giusto et al. 2010) (The percentage shows the distribution of literature in selected research domains). The literature on IoT is mainly focused on sensors, actuators, power sources, i.e.,

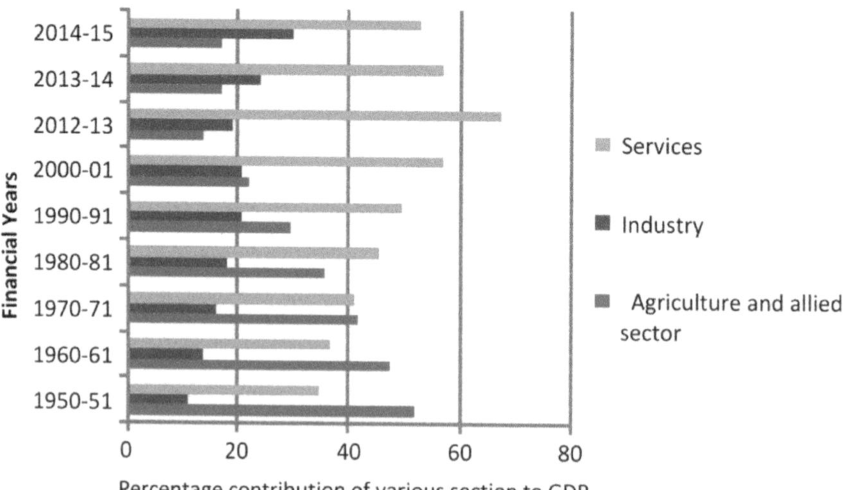


**Fig. 7.1** Sector-wise contribution of GDP of India (1950–2015)

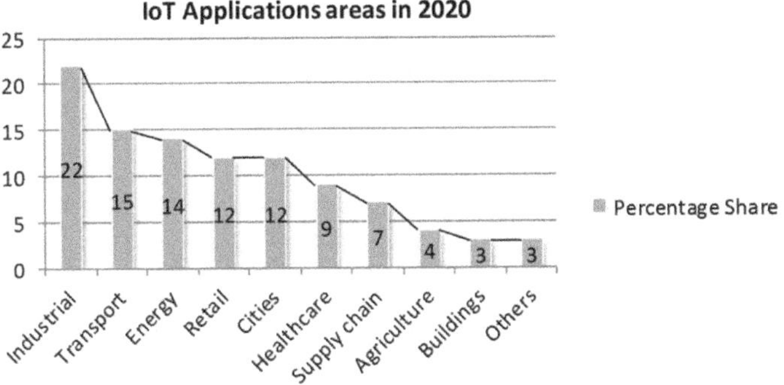


**Fig. 7.2** Global share of IoT projects in different sectors

equipment's and edge computing technologies, communication network, big data computing and storage solutions.

The need of continuous monitoring and control of wide areas implemented in agro-industry and environmental sector makes them suitable domain for IoT-based solutions. It provides several advantages in terms of field automation, data collection, machine learning based on prediction, easy decision-making for owners, seamless planning by managers and policy makers. This technology is applicable at different levels of agro-industry sectors. It can monitor field variables such as soil conditions, irrigation, fertilization, atmospheric conditions, harvesting period and product quality. Also, IoT provides information, connectivity, adaptability and better management to rural farming practices. The low cost, low power electronic devices and computation software with internet connectivity can provide better interaction of human being with physical world. IoT is a vital tool to facilitate beneficiaries such as suppliers, farmers, technicians, business enterprises, distributors, consumers and government policy makers in upcoming years.

The recent thrust area of IoT-enabled smart farming has focused on communication network, energy management, monitoring and logistics. IoT application empowered with Low Power Wide Area Network (LPWAN) technologies such as SigFox, LoRA, narrowband IoT, Bluetooth, Zigbee, WiMax are becoming popular because of energy efficiency, wide coverage and low cost (Barrachina-Muñoz et al. 2017). Since Wireless Sensor Network (WSN) work at wide area and remote locations, it requires multisource energy harvesters and battery free solutions such as solar powered devices or self-powered devices. Thus, energy management of WSN is a crucial area of research for successful implementation of IoT-enabled smart farming. Nowadays, environmental monitoring techniques offer purely autonomous, hostile intervention and resiliency in case of node failure or poor connectivity. This IoT-enabled WSN system can monitor wide field parameters and will improve the overall productivity (Shaikh and Zeadally 2016). The agri-buisness is dependent on food safety and quality control. IoT-enabled logistics framework can resolve the challenges faced by transporters such as perishability and expensive logistics. It creates centralized information system among the food supply chain which can make food storage, transport, e-commerce deliveries seamless and effective (Ruan and Shi 2016). Figure 7.3 shows the basic IoT architecture for smart farming framework. It has four basic layers, physical layers which has WSN network (meant for smart sensing, monitoring and control), communication layer for transmitting the collected data from physical layer to service layer. In this layer, the storage, analysis, visualization of collected data is carried out. The inference is then passed to the application layer which implement the data to monitor, control or predict the environmental and crop conditions.

There are some open challenges and limitations in the adoption of IoT technology in agro-industrial sector. These are briefly pointed below (Talavera et al. 2017).

1. *Standardization/compatibility issue*: Compatibility with legacy infrastructure is an utmost important factor for smooth implementation of IoT. Improved compatibility and standardization among different vendors and security measures for entire IoT-enabled WSN network is one of the limitation for IoT technology adoption.
2. *Energy Management*: Using energy harvesters/alternative power storage modules can increase the life expectancy of electronic smart devices modules.
3. *Data privacy*: End-to-end data privacy and physical integrity is required for IoT-enabled agro-industry and environment.
4. *Cost efficacy*: The per unit cost sharply increases for the total module of high quality sensors, actuators, nodes, internet data access and embedded communication technology.

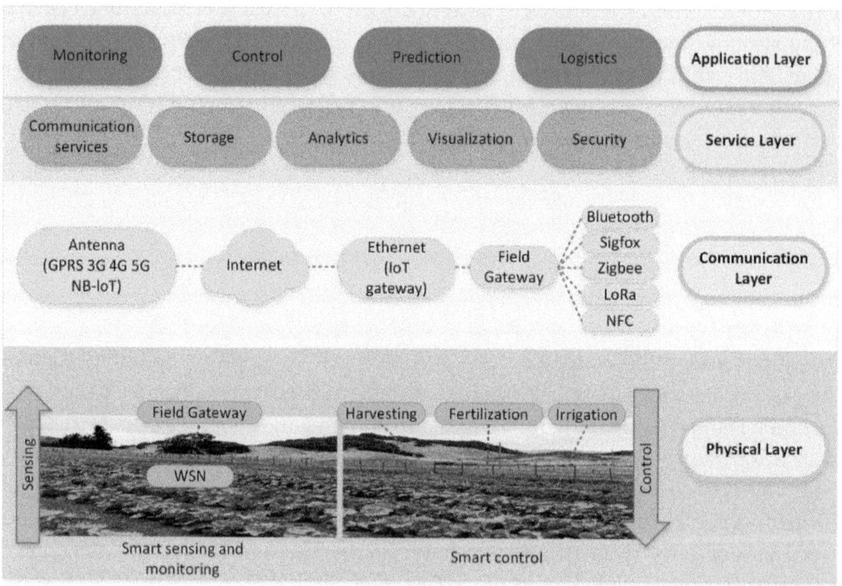


**Fig. 7.3** IoT architecture for smart farming framework

5. *Scalability*: Improved data synchronization and data reliability is the prerequisite for large deployment of IoT technologies.
6. *Software professionals*: The implementation and maintenance of IoT and WSN technology requires skilled technicians/software professionals to refine codes, add features and generate data according to the in-field conditions.

### 7.1.1 Key Drivers of IoT Technology in Smart Farming

IoT-based technology has redefined almost all the sectors including industries, manufacturing, health, energy, climate prediction, digital marketing and agriculture. This revolutionary alteration has created new opportunities along with technological challenges. IoT devices integrated with wireless sensor network and big data computing applications serves various purposes in smart farming such as soil preparation, crop condition, monitoring fertilization, irrigation, pest detection, harvest period prediction and product quality analysis. Thus, it empowers the complete food chain and provides benefits to food industries. In order to feed larger urban community, food production should be doubled by 2050, which creates huge burden on rural community working with traditional agricultural resources and limited cultivable lands.

Considering the traditional farming procedures, 70% of farming time is spent on monitoring and understanding the crop conditions instead of doing proper irrigation, pest detection and fertilization activities. Also, specific crops can be rotated in same field season to season and biologically reach different stages depending on location and temporal difference. To respond to these critical challenges, farmers need remote sensing and communication technologies which can help in producing more yield in less efforts and can continuously observe fields without being present there physically. IoT technology is going to play a huge role in various applications of agricultural domain. They offer capabilities such as remote data acquisition, cloud computing, user interfacing, agriculture process automation, decision-making, logistics and marketing. Such features will modify the agriculture industry. Figure 7.4. summarizes the key drives and challenges faced to implement IoT technology in smart farming (Ayaz et al. 2019).

### 7.1.2 Challenges Faced to Implement IoT Technology in Smart Farming

Engineers and researchers around the world proposed different framework and architecture to implement IoT and big data computing technologies in agricultural sector to meet up the upcoming food crisis. Despite several benefits offered by technological reforms, the major hurdles in implementation of technology in smart farming still

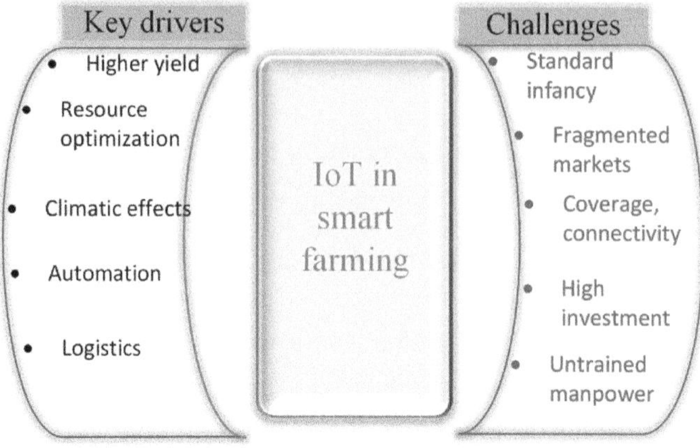


**Fig. 7.4** Key drivers and challenges of IoT technology implementation in smart farming

persists. There are different reasons which create obstacles in successful adoption of IoT technology such as:

- *Standard infancy*: Although several IoT platforms are open source, still their infancy, regulations and governance regarding data privacy, ownership, liabilities and security is still in growth phase (Alam et al. 2017).
- *Fragmented markets*: The compatibility with legacy infrastructure, improved smart devices compatibility among various vendors and data privacy measures need to be completely addressed before in-field application.
- *Coverage and connectivity*: The communication layers offers critical functionalities of interaction between physical layer and service layer. These communication technologies such as Bluetooth, Zigbee, LoRA, SigFox, WiMax offer different coverage and speed of data transfer according to basic characteristics of techniques. To choose the best low power, wide coverage and reliable connectivity communication technology is also a major challenge (Zhang et al. 2019).
- *High investment*: Initially, the investment to implement IoT technology with WSN is very high even for monitoring and control of small agricultural field. Also, the power sources are required to operate the WSN network which will incur cost of batteries for remote monitoring applications. Thus, need of energy harvesters or self-powering smart devices is growing in smart agriculture applications.
- *Untrained manpower*: Unlike the traditional agricultural methods, the IoT-based smart farming requires skilled technicians/skilled professionals to refine features, modify codes and generate desired data.

## 7.2 Major Applications of IoT in Smart Farming

With implementation of IoT technology and wireless sensing network, every aspect of traditional agriculture technique has reformed to gain higher yield, better drought response, land preparation, irrigation, pest control, harvest prediction and production control. The various technological advancements help in different stages of smart farming leading to enhanced overall farming efficiency. The major application of IoT in different stages of agriculture are (Farooq et al. 2019):

- *Soil preparation*: The main goal of soil analysis is to determine the status of soil nutrients, so that appropriate measures can be taken during soil preparation to meet such deficiencies. In order to furnish this detail, comprehensive soil tests are prescribed in different topographical and climatic zones. The soil mapping enables better match of soil properties to different crop varieties. These soil features facilitate precise fertilization, selection of seeds, time to sow and planting depth. Furthermore, multiple crops can be sown in same land making a smarter use of available resources in smart farming. Such soil testing kit tool developed by Agro-Cares facilitate complete lab for soil testing (Accessed 2019). Moderate Resolution Imaging Spectroradiometer (MODIS) sensor can be used to map various soil functional properties to estimate soil degradation risks (Vågen et al. 2016).
- *Irrigation*: Several automated irrigation techniques, such as sprinkler and drip irrigation system, have been opted to reduce water wastage and improve irrigation efficiency unlike flood irrigation method. The crop productivity and quality essentially depend on soil moisture content. The shortage or excess of water may cause reduction in soil nutrients and promote microbial diseases. In order to determine the water necessity of crops, several factors are analyzed such as crop types, soil types, irrigation method, precipitation rate and climatic conditions. Thus, WSN helps in achieving this goal leading to a better productivity efficiency in least physical intervention (Jaiswal and Ballal 2020).
- *Fertilization*: It is a vital step of smart agriculture which helps in precise estimation of required nutrients to be added in the soil to help in crop growth. Fertilization requires site-specific data and depends on several factors like soil types, crop types, fertility rate, absorption rate and weather conditions. Thus, it is not only time consuming but also a quite expensive process. There have been several researches in this domain, NDVI (Normalized Difference Vegetation Index), variable rate technology, geo-mapping, GPS, GPRS and UAVs can be used in IoT-based smart farming.
- *Disease control*: Crop disease control depends on three important components, i.e., sensing system, analyzing system and treatment. The disease detection and treatment is possible by image processing of crop area using field sensors, UAVs or remote sensing satellites. Remote sensing image covers wide range area and offers high efficiency in low cost setup. Field sensors on the other hand offers more functionalities and is expensive in comparison to remote imaging techniques.
- *Crop monitoring and estimation of harvest period*: Crop productivity monitoring is an essential stage of smart farming as it plays important role in crop yield

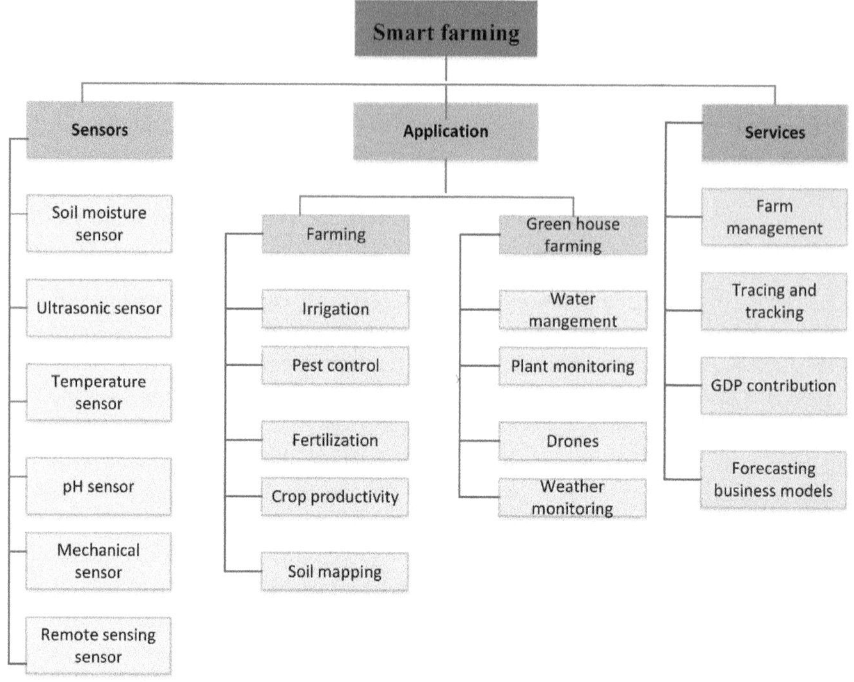


**Fig. 7.5** Block diagram of sensors, applications and services in smart farming

estimation. The productivity is largely affected by pollination with good quality pollen grains and climatic conditions. The yield quality and crop maturity are essential parameters which decides the right time of harvesting. Forecasting harvesting period not only maximize the total benefit to farmers but also provide an opportunity to prepare for storage and logistic beforehand.

- *Green-house farming*: Green-house farming is the oldest type of agriculture where indoor grown crops are least affected by outside environment. Thus, the growth of crop is dominated by controlled environment inside such as structure of shed, covering material to protect from heavy wind, ventilation, moisture monitoring and decision-making technique. An IoT-based prototype is depicted for green-house farming to control humidity, temperature and light and is done using MicaZ nodes (Akkaş and Sokullu 2017). Figure 7.5 shows the block diagram of sensor, applications and services in smart farming.

## 7.3 Equipment and Technologies for Smart Farming

Every domain of agriculture that can be digitalized, automated and controlled will be profited with the implementation of IoT technologies and solutions. Based on this fact, several efforts are made to invent more sophisticated tools such as UAVs and robots that can perform range of operations such as irrigation, weeding, fertilization, harvesting and logistics. The successful implementation of smart farming depends on the accuracy of data collected from two ways, i.e., first through imaging devices with remote sensing satellites and UAVs and second from in-field-based sensors. The collected data can be stamped with precise location information by GPS so that site-specific planning and measure can be carried out for any specific problems.

### 7.3.1 Wireless Sensors

Wireless sensors are crucial part of smart farming to collect crop condition information. They can be used in standalone mode or further be integrated with advanced data computing tools depending upon application requirements (Navulur et al. 2017).

- *Acoustic sensors*: It performs several purposes including soil preparation, weeding and crop harvesting. It is a portable and low cost solution catering to fast response data transfer (Srivastava et al. 2013).
- *Airflow sensors*: These are capable of accurate measurement of soil permeability and moisture, which helps in distinguishing different types of soil.
- *Electrochemical sensors*: These sensors are used to assess the significant soil characteristics to analyze the soil nutrient level and pH value (Cocovi-Solberg et al. 2014).
- *Optical sensor*: The working principle of these sensors are based on light reflectance phenomena and helps to detect soil organic content, color, presence of soil nutrients.
- *Ultrasonic sensors*: These sensors are economic and have a lot of application in wide areas. They are used in smart farming for tank level monitoring, uniform spray coverage, monitoring crop canopy and weed detection.
- *Optoelectric sensors*: These sensors differentiate in plant type, detect weeds, herbicides and help in better crop productivity (Pajares 2011).
- *Mass-flow sensors*: These sensors are used in crop harvesting and crop yield monitoring as it can measure grain flow when crop passes through harvesters.
- *Mechanical sensor*: These sensors assess soil mechanical properties to indicate the variable level of compaction. These sensor cut through the soil and record the force measured by strain guage.
- *Soil moisture sensors*: These sensors are applicable to categorize hydrological behavior such as flow and water level. It is mostly used to measure water presence in soil, rainfall and stream flow.
- *Remote sensing*: It captures, stores and transmits geographic and spatial data. Agros sensor is one of the leading satellite-based sensor used to collect, process and analyze environmental data from mobile communication platforms worldwide (Rose and Welsh 2010).

### 7.3.2 *Communication Technologies*

Communication and data logging are two major functionalities of precision farming. It should be secure, reliable and provide wide coverage connectivity among various smart devices. To achieve the communication reliability, Ethernet, telecom operators and WiFi can play a vital role in agro-industry sector. Based on the availability, application requirement, communication mode and scalability, different communication protocols can be opted.

- *IEEE 802.11 WiFi*: It can accommodate several standards such as IEEE 802.11 a, 802.11b, 802.11n operating at different bandwidths, i.e., 5 GHz, 2.2 GHz and 60 GHz, respectively. Data transfer rate can vary between 1 Mb/s and 7 Gb/s. Coverage range is in-between 20 and 100 m (Xiao et al. 2006).
- *WiMax*: It provides broadband multi-access connectivity including mobile communication through wired or wireless connections. It operates in the range of 1.5 Mb/s to 1 Gb/s.
- *LoRA WAN*: LoRA wide area network is developed by TM alliance and is open source protocol. It assures interoperability between multiple operators specially designed to improve crop yield.
- *Mobile communication*: There exists multiple generation for mobile communication, i.e., 2G/3G/4G/5G. IoT devices communicate using these mobile communication network to monitor field data such as soil, crop growth and climatic conditions (Feng et al. 2019).
- *RFID*: It works on the principle of assigning unique number individually to each smart devices in order to record information. It can be used for low cost smart farming environment for receiving and transmitting sensor information.
- *SigFox*: SigFox is a narrowband wireless cellular network which has low data rate suitable for IoT and M2M communication (Pitì et al. 2017).
- *Bluetooth*: It is suitable for low power and low range personal area network suitable for short range communication (Ruiz-Garcia et al. 2009).
- *Zigbee*: This is used for device-to-device communication with low power data rates. It helps smart farming environment by establishing low cost, bidirectional communication between IoT devices and remote servers (Ray 2017).

Table 7.1 gives the comparison between the above-mentioned communications technologies based on their features.

## 7.4 Role of Big Data in Smart Farming

The fundamental concepts of IoT and device-to-device communication was stated by Kevin Aston in 1999 (Ashton 2009). IoT is versed as convergence of three visions: things oriented, internet oriented and semantic oriented. It helps in real-time information sharing between wireless networks. IoT generates a tremendous volume of

**Table 7.1** Comparison of wireless communication technologies used in smart farming

| Parameters | Standard | Frequency band | Data rate | Coverage | Energy efficiency | Cost |
|---|---|---|---|---|---|---|
| WiFi | IEEE 802.11 | 5–60 GHz | 1 Mb/s to 7 Gb/s | 20–100 m | Low | High |
| LoRA | LoRA WAN R1.0 | 868/900 MHz | 0.3–50 Kb/s | <30 km | Very high | High |
| WiMAX | IEEE 802.16 | 2Ghz to 66 GHz | 1 Mb/s to 1 Gb/s | <50 km | Medium | High |
| Mobile communication | 2G-GSM, CDMA, 3G, 4G-LTE, 5G | 865 MHz to 2.4 GHz | 40–250 Kb/s | Cellular area | Medium | Medium |
| RFID | ISO 18000-6C | 860–960 MHz | 40–160 kbits/s | 1–5 m | High | Low |
| Zigbee | IEEE 802.15.4 | 2.4 GHz | 20–250 Kb/s | 10–20 m | High | Low |
| SigFox | SigFox | 200 kHz | 100–600 bit/s | 30–50 m | High | Low |
| Bluetooth | IEEE 802.15.124 GHz | 24 GHz | 1–24 Mb/s | 8–10 m | Very high | Low |

data called as big data which requires recent trends solutions for data management to provide process insights and decision-making. Big data has different characteristics such as high volume, high variety, veracity and high velocity. IoT framework facilitate with ubiquitous network of WSN, data sources, smart devices with a potential to incorporate existing applications and give a deeper insight for future upgradation and solutions. The exponential growth in variety of sensor and generated data sets makes it complex and difficult for user to extract knowledge by analyzing the data and share it on the cloud. The solution to this issue can be devised by integrating IoT and big data computing environment. Big data tools can collect, analyze and evaluate high volume of data and extract useful information using data mining technique.

Big data technologies in smart agriculture goes beyond the crop production, it affects the entire food chain by transforming the agro-industrial environment. It provides deeper and accurate predictive insight in farming operations, analyzes volume of real-time data to extract decision and redesign business models which can revolutionize food markets. There are several key drivers of development of big data technology in smart farming such as farm process, farm management, data chain and network management. Data chain forms an integral part of big data applications. It refers to sequence of functions originating from data capture to decision-making and data marketing. Data chain consists of technical layer that gathers the raw data and convert it into information and passes it to business layer that makes decision and provides data services to owners. The technical layer and business layer is inter related to form data value chain. Figure 7.6 presents the Schematic diagram of functions performed in data chain for big data applications (Dumbill 2014).

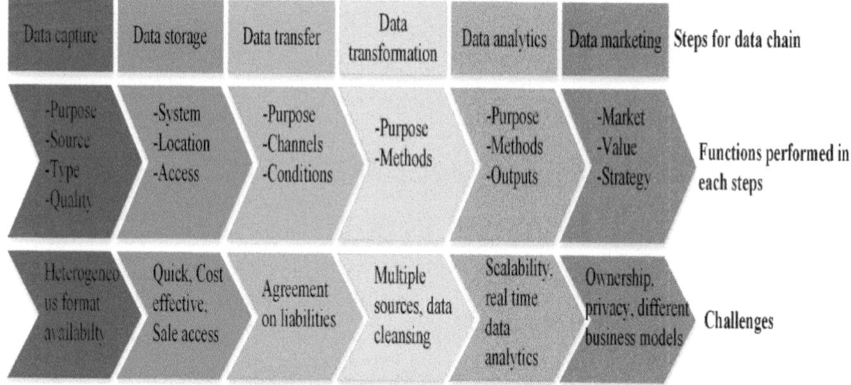


**Fig. 7.6** Data chain functions and challenges faced in Big data applications

### 7.4.1 *Big Data Tools for IoT Applications*

There are several tools mentioned in the literature which support big data computing for IoT applications. Some of them are discussed below (Hajjaji et al. 2021).

- *Hadoop*: Apache foundation facilitating several open source projects out of Hadoop serves the purpose of distributed storage and processing of massive datasets enabled with big data architecture. The core of Apache Hadoop consists of two subprojects, HDFC for storage and MapReduce for processing.
- *Apache Flume*: This platform provides efficient distribution services to log file and events into Hadoop.
- *Kafka*: It is a distributed, highly efficient and publish subscribing messaging platform. It is used for large-scale message processing and enables data to be implemented in various application software.
- *Spark and Strom*: These open source framework are produced by Apache to analyze massive data like Hadoop. Spark is faster in computation because of its memory capabilities stream processing in comparison with Hadoop.
- *Apache Hive*: Hive utilizes SQL-based interface to poll the data stored at different files and databased to run at the top level of Hadoop.
- *NoSQL database*: As name suggests, not only SQL database system manages large volume of data of unstructured and semi-structured characteristics perfectly which was not earlier possible with traditional databases.
- *Cloud computing*: It is a computing technique which relies on remote server to analyze a large volume of data for fast speed data analysis. Cloud computing ensures the basic layer of computing resources and backing higher layer of big data processing.

### 7.4.2 Benefits of Integrating IoT with Big Data Tools

There are several benefits of integrating both the recent techniques together to collect large amount of data from smart devices using IoT and analyze and store it using big data tools.

- *Connectivity*: It is the most crucial factor in computing and information exchange by aggregating massive volume of data from multiple sources. The communication network between interlinked nodes and data management system (DMS) caters as the backbone of the complete application layers. Thus, a variety of low cost, wide coverage and efficient communication technologies are now available with distinct communication protocols and security measures to fit the need of smart farming environment (Babar and Arif 2019).
- *Collection of heterogeneous data from multi-source system*: By integration of IoT technology with data mining techniques, multiple sources such as sensors, actuators and other smart devices can deliver useful information and deeper insight to operational modes. Such knowledge can be used to form predictive models (Bandyopadhyay et al. 2011).
- *Data storage*: Storage of unstructured and heterogeneous data using data management system is a challenging task. In order to resolve this issue, big data technologies such as Hadoop, Spark, Strom, No SQL databases have been introduced which helps in collection, storage and preservation of large volume of data (Tu et al. 2020).
- *Data analysis*: Efficient, real-time processing software solutions and technological capabilities enabled processors are required for accurate data analysis for dynamic and demanding situations in order to extract useful information for decision-making and control.
- *Cost efficacy*: Big data technologies are mostly open sourced and offer low cost solution for development and deployment of new applications. It makes implementation of IoT integrated with big data tools easier to be opted by developing countries to benefit rural community (Wolfert et al. 2017).

### 7.4.3 Key Challenges in Implementation of IoT and Big Data Tools Applications

In order to successfully implement the IoT and big data technologies to upgrade the agricultural environment, the following hurdles need to be discussed and resolved.

- *Security and privacy*: This is a major challenge in implementing IoT since the risk of spoofing, data hacking, manipulation and cyber-attack on confidential information can affect the performance of big data technologies and may result in faulty decision-making (Asghari et al. 2019).

- *Massive volume*: In order to deal with massive volume heterogeneous data, the operational mechanism need to be upgraded with latest technologies for data collection, storage, processing and management.
- *Velocity*: The speed of data access and generation is a crucial issue which also generates a concern for real-time dynamic data analysis.
- *Veracity and variety*: The quality, accuracy and applicability of data is an important concern. The data should be free of noise, biases and abnormalities. The data can be of different variety; structured, semi-structured and unstructured which makes it difficult to organize (Ardagna et al. 2018).
- *Visualization and value*: The visualization technique such as graph and charts should be effectively chosen to highlight the inferences drawn from the data. Also, the amount of data extracted for decision-making decides its cost and benefit ratio. The data which is important only need to be extracted and valued.
- *Knowledge extraction*: Process of extraction of knowledge from heterogeneous data collected from multi-source system having different structures is a difficult task.

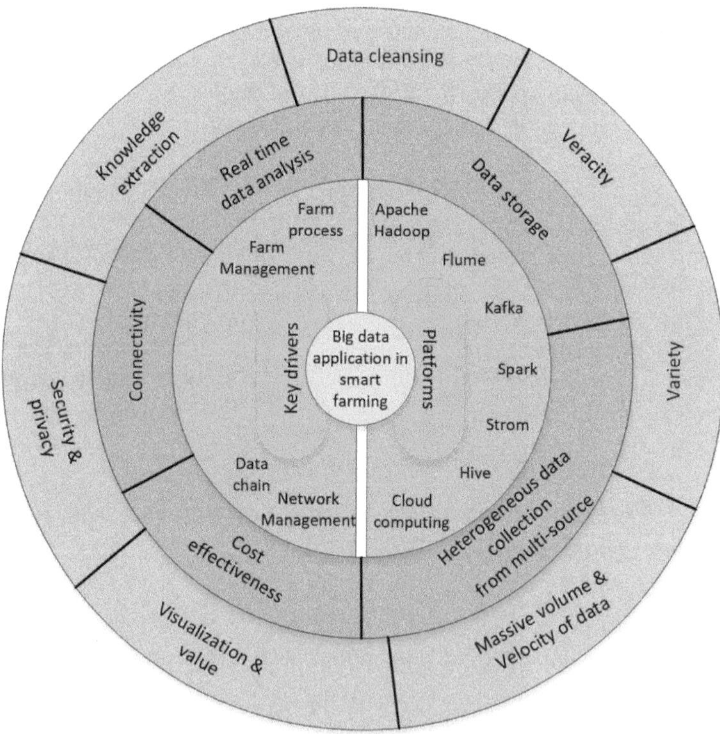


**Fig. 7.7** Benefits and challenges of Big data applications in smart farming

Figure 7.7 presents the summary of different aspects related to big data applications.

## 7.5 Future Opportunities in the Domain of Smart Farming

Researchers around the world are working to reduce the overall cost of hardware and software deployed in IoT applications to farming to maximize the crop yield efficiency. The standardization in IoT application and platforms is important to clear the interoperability and compatibility hurdles faced by service providers and active users. Energy management is also a most crucial issue for IoT-based system implementation. Green computing technologies need to be integrated with IoT, so that the smart devices consumes less amount of power and may increase system life expectancy. Fault tolerance is almost an unaddressed topic in concern of IoT-based smart farming researches. To make flawless system, fault tolerance level of the system should be kept at high priority. Hardware modules may fail due to depleted batteries or other reasons, can incur heavy maintenance cost.

Machine learning algorithms and intelligent techniques integrated with IoT technology may help to access the predictive and behavioral analysis of smart farming activities. Portability of the IoT architecture need to be enhanced to increase its usability in real-time environment. Cloud-enabled computing, storage systems and agro-logistic system are recent research topics that can help to leverage high profit margins to the farmers and make smart farming a seamless experience. Big data analytics and cloud computing web services may be developed to facilitate the knowledge of farmers about the smart farming. Furthermore, government and policy makers should come together to pave new policy for improving agriculture-related big data analytics.

## References

Accessed: Apr. 15, 2019. [Online]. https://www.agrocares.com/en/products/lab-in-the-box/

Akkaş MA, Sokullu R (2017) An IoT-based greenhouse monitoring system with Micaz motes. Procedia Comput Sci 113:603–608

Alam F, Mehmood R, Katib I, Albogami NN, Albeshri A (2017) Data fusion and IoT for smart ubiquitous environments: a survey. IEEE Access 5:9533–9554

Ardagna D, Cappiello C, Samá W, Vitali M (2018) Context-aware data quality assessment for big data. Futur Gener Comput Syst 89:548–562

Asghari P, Rahmani AM, Seyyed Javadi HH. Internet of things applications: a systematic review. Comput Netw 148:241–261

Ashton K (2009) That 'internet of things' thing. RFID J 22(7):97–114

Ayaz M, Ammad-Uddin M, Sharif Z, Mansour A, Aggoune E-H (2019) Internet-of-things (IoT)-based smart agriculture: toward making the fields talk. IEEE Access 7:129551–129583

Babar M, Arif F (2019) Real-time data processing scheme using big data analytics in internet of things based smart transportation environment. J Ambient Intell Humaniz Comput 10:4167–4177

Bandyopadhyay S, Sengupta M, Maiti S, Dutta S (2011) Role of middleware for Internet of things: a study. Int J Comput Sci Eng Surv 2(3):94–105. https://doi.org/10.5121/ijcses.2011.2307

Barrachina-Muñoz S, Bellalta B, Adame T, Bel A (2017) Multi-hop communication in the uplink for LPWANs. Comput Netw 123:153–168. https://doi.org/10.1016/j.comnet.2017.05.020

Cocovi-Solberg DJ, Rosende M, Miró M (2014) Automatic kinetic bioaccessibility assay of lead in soil environments using flow-through microdialysis as a front end to electrothermal atomic absorption spectrometry. Environ Sci Technol 48(11):6282–6290

Dumbill E (2014) Understanding the data value chain. IBM Big Data & Anal Hub 10

Farooq MS, Riaz S, Abid A, Abid K, Naeem MA (2019) A survey on the role of IoT in agriculture for the implementation of smart farming. IEEE Access 7:156237–156271

Feng X, Yan F, Liu X (2019) Study of wireless communication technologies on Internet of Things for precision agriculture. Wireless Pers Commun 108(3):1785–1802

Giusto D, Iera A, Morabito G, Atzori L (eds) (2010) The internet of things. In: 20th Tyrrhenian workshop on digital communications. Springer, ISBN 978-1-4419-1673-0

Hajjaji Y, Boulila W, Farah IR, Romdhani I, Hussain A. Big data and IoT-based applications in smart environments: a systematic review. Comput Sci Rev 39:100318

Jaiswal S, Ballal MS (2020) Fuzzy inference based irrigation controller for agricultural demand side management. Comput Electron Agric 175:105537

Navulur S, Giri Prasad MN (2017) Agricultural management through wireless sensors and internet of things. Int J Electr Comput Eng 7(6):3492

Pajares G (2011) Advances in sensors applied to agriculture and forestry, 8930–8932

Pitì A, Verticale G, Rottondi C, Capone A, Schiavo LL (2017) The role of smart meters in enabling real-time energy services for households: The Italian case. Energies 10(2):199

Ray PP (2017) Internet of things for smart agriculture: technologies, practices and future direction. J Ambient Intell Smart Environ 9(4):395–420

Rose I, Welsh M (2010) Mapping the urban wireless landscape with Argos. In: Proceedings of the 8th ACM conference on embedded networked sensor systems, pp 323–336

Ruan J, Shi Y (2016) Monitoring and assessing fruit freshness in IoT-based ecommerce delivery using scenario analysis and interval number approaches. Inf Sci 373:557–570. https://doi.org/10.1016/j.ins.2016.07.014

Ruiz-Garcia L, Lunadei L, Barreiro P, Robla I (2009) A review of wireless sensor technologies and applications in agriculture and food industry: state of the art and current trends. Sensors 9 (6):4728–4750

Shaikh FK, Zeadally S (2016) Energy harvesting in wireless sensor networks: a comprehensive review. Renew Sustain Energy Rev 55:1041–1054. https://doi.org/10.1016/j.rser.2015.11.010

Srbinovska M, Gavrovski C, Dimcev V, Krkoleva A, Borozan V (2015) Environmental parameters monitoring in precision agriculture using wireless sensor networks. J Clean Prod 88:297–307. https://doi.org/10.1016/j.jclepro.2014.04.036

Srivastava N, Chopra G, Jain P, Khatter B (2013) Pest monitor and control system using wireless sensor network with special reference to acoustic device wireless sensor. In: International conference on electrical and electronics engineering, vol 27

Talavera JM, Tobón LE, Gómez JA, Culman MA, Aranda JM, Parra DT, Quiroz LA, Hoyos A, Garreta LE (2017) Review of IoT applications in agro-industrial and environmental fields. Comput Electron Agric 142:283–297

Tu L, Liu S, Wang Y, Zhang C, Li P (2020) An optimized cluster storage method for real-time big data in internet of things. J Supercomput 76(7):5175–5191

Vågen T-G, Winowiecki LA, Tondoh JE, Desta LT, Gumbricht T (2016) Mapping of soil properties and land degradation risk in Africa using MODIS reflectance. Geoderma 263:216–225

Wagh R, Dongre AP (2016) Agricultural sector: status, challenges and it's role in Indian economy. J Commer Manag Thought 7(2):209

Wolfert S, Ge L, Verdouw C, Bogaardt M-J (2017) Big data in smart farming—a review. Agric Syst 153:69–80

Xiao Y, Chen H-H, Sun B, Wang R, Sethi S (2006) MAC security and security overhead analysis in the IEEE 802.15. 4 wireless sensor networks. EURASIP J Wirel Commun Netw 2006:1–12
Zhang A, Liu S, Sun G et al (2019) Clustering of remote sensing imagery using a social recognition-based multi-objective gravitational search algorithm. Cogn Comput 11:789–798

# Chapter 8
# Fermat Point-Based Wireless Sensor Networks: A Default Choice for Measuring and Reporting Farm Parameters in Precision Agriculture

**Kaushik Ghosh and Sugandha Sharma**

**Abstract** The domain of precision and smart agriculture has been influenced by wireless sensor networks for better when it comes to cost optimization in terms of manpower. Deployment of sensor nodes in agricultural lands for monitoring different crucial farm parameters has been a very prominent area of interest to the research fraternity for a decade and half. In this chapter, we have advocated for installing a WSN-based framework that works on the principle of Fermat point for measuring farm parameters and reporting them to different monitoring points. It has been shown here that of the different data forwarding techniques, Fermat point-based techniques outperform their counterparts in terms of energy efficiency. The enhanced lifetime thereby obtained reduces the running cost of the installation.

**Keywords** Precision agriculture · Wireless sensor network · Energy efficiency · Fermat point

## 8.1 Introduction

Of the different application domains where Wireless Sensor Networks (WSNs) fit in, precision agriculture (PA) is a prominent one. Introducing WSN in precision and smart agriculture has had a host of positive effects in the form of reducing the manpower requirements and reduction in network installation cost. Incorporating technology in agriculture is nothing new. However, the advent of wireless technology was able to eliminate up to 80% of the cost introduced as a result of wiring (Wang et al. 2006). Wireless sensor nodes installed in PA acquire measurements for different farm parameters and report the same to a sink or collector point. In many other cases, a node plays the role of data aggregator as well. The accumulated data are then forwarded to the collector points (Wark et al. 2007). In either case, having more than one sink is rather a necessity due to the reasons mentioned below:

(i) even load distribution for data collection
(ii) reducing the extent of sinkhole problem

K. Ghosh (✉) · S. Sharma
School of Computer Science, UPES, Dehradun, Uttrakhand, India



A. Choudhury et al. (eds.), *Smart Agriculture Automation using Advanced Technologies*, Transactions on Computer Systems and Networks,
https://doi.org/10.1007/978-981-16-6124-2_8

(iii) reduction in the distance covered by a packet in terms of the number of hops encountered
(iv) data collection over multiple interfaces
(v) increasing network lifetime
(vi) removing single point failure.

Energy optimization and network lifetime enhancement in Wireless Sensor Network (WSN) are very common research domains, and quite a handful of energy efficient routing protocols were proposed in the past two decades. Although most of the proposed protocols are for networks with a single sink, a few of them have been proposed for multiple sinks as well. It is beneficial to have multiple, redundant sinks in a WSN due to the reasons stated above.

When it comes to forwarding the same packets to multiple sinks, Fermat point-based approach for forwarding data is most sought for in order to achieve energy efficiency and enhanced network lifetime. This technique guarantees to maximize network lifetime by reducing energy consumption to minimum. This becomes possible as this technique ensures that the packets are transmitted from source to different sinks by traveling minimum distance (Ssu et al. 2009).

In this chapter, we have, therefore, discussed an energy-efficient Fermat point-based framework of WSN, which may be used for measuring different farm parameters and reporting them to multiple sinks. One primary deliverable of a WSN-based framework is to ensure maximum operation time of the installed setup through lifetime maximization of the network. Lifetime maximization in a WSN not only amortizes the installation cost but also ensures a hassle-free operation period of the setup for a prolonged duration of time.

The rest of the paper is organized as follows: in Sect. 8.2, we have discussed some of the related works in the field precession agriculture involving WSNs. In Sect. 8.3, we discussed the proposed Fermat point-based framework. Section 8.4 contains the lifetime comparison of some Fermat point-based protocols with their non-Fermat point-based counterparts and Sect. 8.5 contains the conclusions.

## 8.2 Literature Survey

Kassim et al. (2014) listed down different applications of WSN in precession agriculture. Their work focused on how WSN may become useful in decision-making, resources optimization, farming and land monitoring in a precession agricultural framework. Their work emphasized how real-time farm information may be obtained using WSN as backbone and how the same framework can be converted to a IoT-based system, through incorporation of Internet. A precision irrigation system through software process control was discussed in the paper.

A review on WSN-based machine learning techniques for precision agriculture was discussed by Mekonnen et al. (2019) and proposed a method for handling big data with spatial and temporal variations of multiple modalities. The work discusses different applications of machine learning in analyzing sensor data in the smart agriculture ecosystem. The authors argue that machine learning techniques when added to WSN make it possible to extract necessary farm information and thereby providing deeper insights into the received data.

Sahitya et al. (2016) stated how precision agriculture can reduce the requirement of manpower and thereby reducing the cost. They proposed a wireless sensor network-based system for smart agriculture to give the right inputs to the crops. The Arduino-based sensors measure farm parameters and send to a single sink. The actuators are activated after analyzing the data at sink.

A real-life case study of an irrigation management system (IMS) through WSN was exhibited (Mafuta et al. 2013). The work described a fully automated and low-cost irrigation management system that is suitable as per the socio-economic strata of small farmers in developing countries. To make the system self-sufficient in terms of power, rechargeable solar photovoltaic batteries were used for all electrical devices. The deployed system reported not only farm parameters like soil moisture and temperature but also network parameters like WSN link performance, and electrical parameters like battery power levels to the sink through a GPRS-based system.

A cross-layer approach was taken by Sahota et al. (2011) for collecting periodic data from fixed locations in an agricultural land. The scheme proposed multiple power modes for the nodes at the physical layer, to achieve energy efficiency. The MAC protocol designed synchronizes the sender and receiver during packet forwarding. In the network layer, they proposed an energy-efficient routing scheme based on a routing tree.

Ghosh et al. (2017) proposed an energy regulating framework for recording soil moisture and reporting the same to three different sinks. The protocol emphasized on energy efficiency, and the results showed that for grid deployment of the nodes, their proposed scheme performed way better as compared with protocols like LEACH and TEEN for measuring and reporting farm parameters.

## 8.3 Proposed Framework

The proposed framework will comprise multiple sensor nodes and around three or four sink nodes. The nodes will be deployed throughout the farm in a grid fashion. The nodes deployed will be equipped with multiple sensors, which are capable of measuring different farm parameters like soil moisture, soil temperature, salinity in the soil, pH, etc. The sensed data will be forwarded to the sinks only if the value of a sensed parameter is not within the permissible thresholds.

The grid deployed sensors will report the required farm parameters to multiple sinks. The reasons for having multiple sinks in the framework are threefold: (i) reduction in sinkhole problem, (ii) providing infrastructure support over multiple interfaces and (iii) eliminating single point failure.

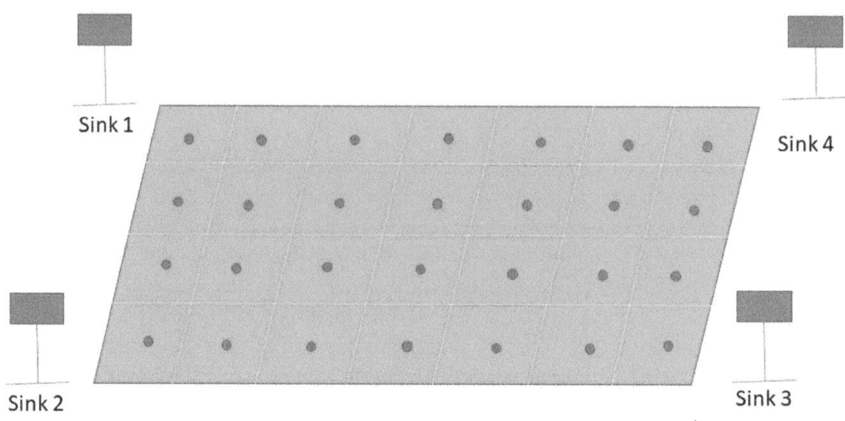


**Fig. 8.1** Sensor nodes deployed in grid fashion over the farm

The nodes deployed in a grid fashion over a rectangular farmland and having four sinks in its four corners will look somewhat as given in Fig. 8.1.

The nodes will transmit packets to the sinks in multiple hops using a data forwarding scheme that works on the principle of Fermat point. Fermat point is a sole point located inside a polygon, such that the sum of the distance of all the vertex of the polygon from it is minimum in comparison to any other point inside the bounds of the polygon.

Ghosh et al. (2015) has shown in their work that Fermat point-based data forwarding technique is most suitable for a multi-sink environment. This is because, through Fermat point-based data forwarding in a multi-sink scenario, it is possible to minimize the total physical distance a packet needs to travel. The distance traveled by a packet is the most important determining factor behind energy consumption in a WSN and can be seen from the radio models proposed by different researchers in their works during the course of time (Heinzelman et al. 2002; Hwang and Pang 2007; Ghosh et al. 2016).

As per our knowledge, a Fermat point-based scheme was first taken into consideration for data forwarding in WSN by Lee and Ko (2006). Thereafter, it became a default choice of the researchers for reducing the packet traveling distance in scenarios comprising of static nodes. Figure 8.2 gives the Global Minima Scheme used (Ghosh et al. 2009) for determining the Fermat point of a regular polygon of n sides (Table 8.1).

After a source node detects the Fermat point, it passes on data packet to the node nearest to Fermat point for data forwarding, in multiple hops. Let us label this node as the Fermat Node (FN) for that particular source. Thereafter, The FN will forward the same packet to all the n sinks, in multiple hops. However, before FN can forward the packet, it would aggregate the packets. The degree of aggregation is determined through the value of AGFACT. AGFACT = 1 signifies no aggregation and AGFACT = $m$ would signify aggregation of m packets. During packet forwarding, a node will select one of its neighbors based on a parameter called the Forwarding

**Minima Algorithm**

Input: Coordinates of the sender node and that of different geocast regions.
Output: (fx, fy): Coordinates of the Fermat_Point.
Tdist : Total distance traveled by the packet.
Tpow: The sum of power consumed by all the intermediate nodes to forward the packet to m geocast regions

```
max_x=MAX_x (Sx, GRX(N));
max_y=MAX_y (Sy, GRY(N));
min_x=MIN_x (Sx, GRX(N));
min_y=MIN_y (Sx, GRY(N));
dx=0;  /*          Initialize dx */
dy=0    /* Initialize dy */
flag=0; /* To check the Fermat Point*/
for ( i=min_x; i<max_x; i++ )
{ if (flag==1) break; /*  Fermat point found  */
for ( j=min_y; j=max_y; j++ )
{ x=i
y=j
for ( k=0;k<n;k++ )
{ dx+= termdx ( x,GRX(k),y,GRY(k) );
dy+= termdy ( y,GRY(k),x,GRX(k) );
} /* end of for loop (line-11)*/
if ( dx==0 && dy==0 )
{ flag=1; /* Fermat point found */
break
}
dX=0; dY=0;
} /* end of for loop (line-8) */
}/* end of for loop (line - 6) */
if( flag==1 )
{ fx=x; fy=y; } /* Fermat point */
Tdist= Total_Dist ( Sx, Sy, GRX(N), GRY(N), fx, fy );
Tpow= Total_Pow ( Tdist )
```

**Fig. 8.2** Global minima scheme of finding Fermat point

Potential, denoted by the symbol κ. Forwarding potential of a neighbor node a, for some destination/sink node b is calculated as

$$\kappa_{ab} = \text{res\_energy}_a / \text{dist}_b \tag{8.1}$$

wherein,

$\text{res\_energy}_a$ = residual battery power of node a in milli-Joules.

**Table 8.1** Variables used in global minima algorithm

| Symbols | Definitions |
|---|---|
| max_$x$ | Max value of $X$ |
| max_$y$ | Max value of $Y$ |
| min_$x$ | Min value of $X$ |
| min_$y$ | Min value of $Y$ |
| $(S_x, S_y)$ | Sender's coordinates |
| $n$ | Total number of geocast regions |
| GRX[] | Array containing the coordinate values of $X$ for all the geocast regions |
| GRY[] | Array containing the coordinate values of $Y$ for all the geocast regions |
| d$x$ | Minima value of $X$ |
| d$y$ | Minima value of $Y$ |
| $(f_x, f_y)$ | Fermat point's coordinates |
| $T_{\text{dist}}$ | Total transmission distance |
| $T_{\text{pow}}$ | Total power spent in transmitting a packet from the sender to all the geocast regions |

$\text{dist}_\text{b}$ = distance of node a from a sink b.

So, the neighbor with the highest value of κ will be forwarded the data packet until the packet reaches the destination (FN or sink(s)).

A workflow diagram for the entire process is given in Fig. 8.3.

## 8.4 Results

The results show how a Fermat point-based scheme scores over non-Fermat point-based data forwarding techniques.

Figure 8.4 compares the performance of the global minima scheme with that of greedy forwarding scheme. It shows that with an increasing number of sinks, the power consumption in greedy forwarding technique increases much more rapidly as compared with its Fermat point-based counterpart. The results were taken following the 1 out of n definition of network lifetime. It states that a network is considered alive as long as at least one of the nodes is alive (Madan et al. 2006).

Another result that considered the m out of n definition (Dietrich and Dressler 2009) of network lifetime is shown in Fig. 8.5. As per this definition, the sensor network will be considered alive, till $m$ nodes out of n are alive ($m < n$).

In Fig. 8.5, we have taken a data forwarding scheme using principles of Fermat point, which forwards packet based on the forwarding potential of a node (Ghosh and Das 2013). Forwarding potential is the ratio of residual energy of the node to its distance from the sink. The proposed scheme was compared with residual energy-based forwarding and greedy forwarding. The network was considered alive till 75% of the total nodes are alive, that is, the results were taken till 25% of the nodes faded

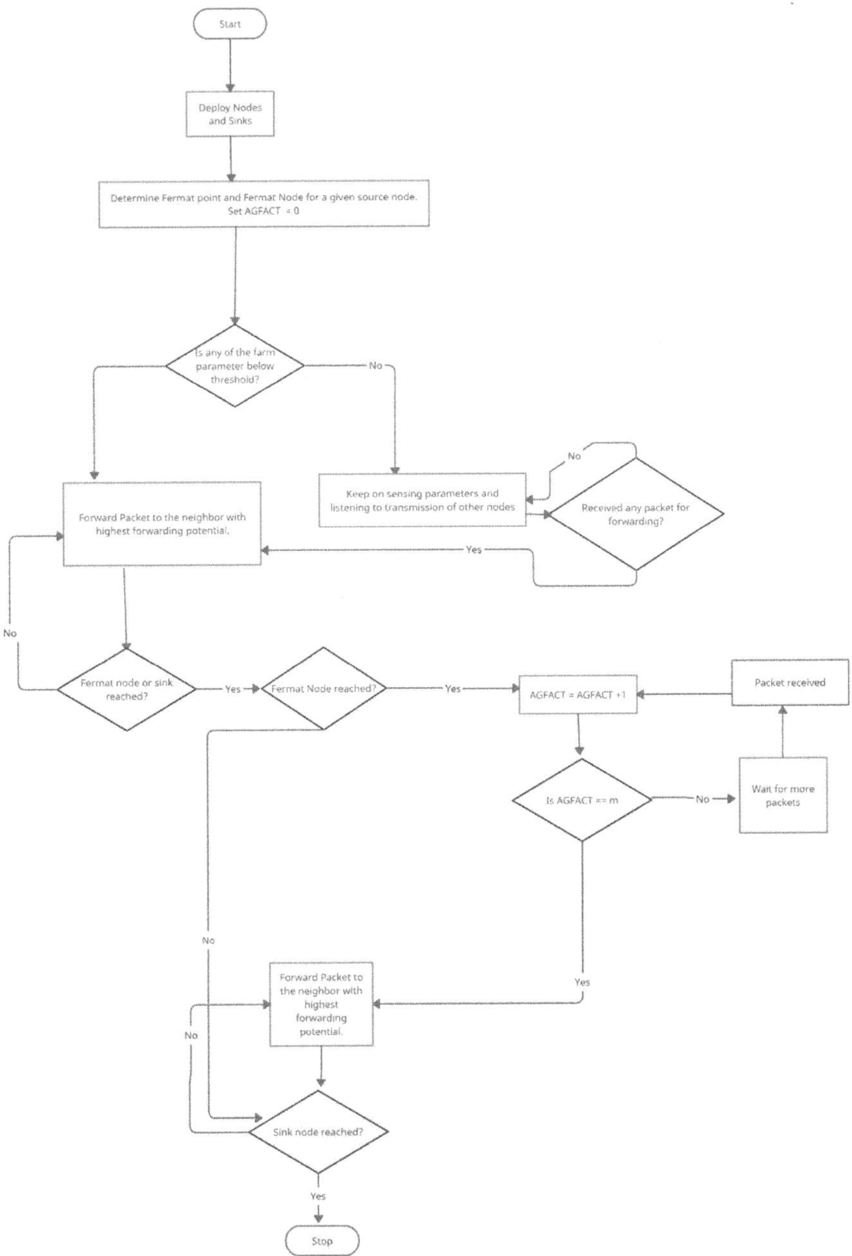


**Fig. 8.3** Flowchart for the process of sensing and packet forwarding

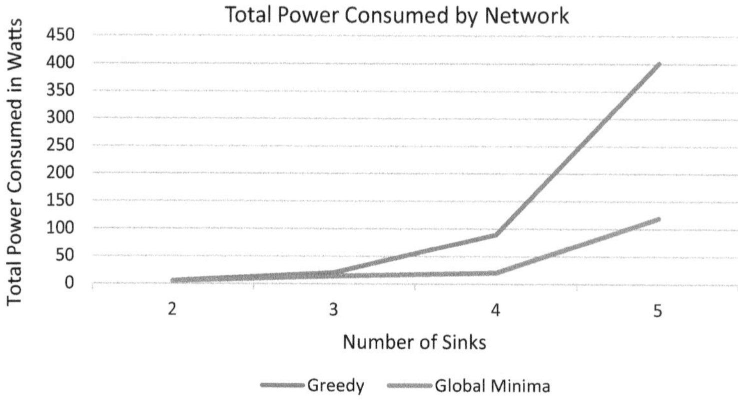


**Fig. 8.4** Power consumed by network

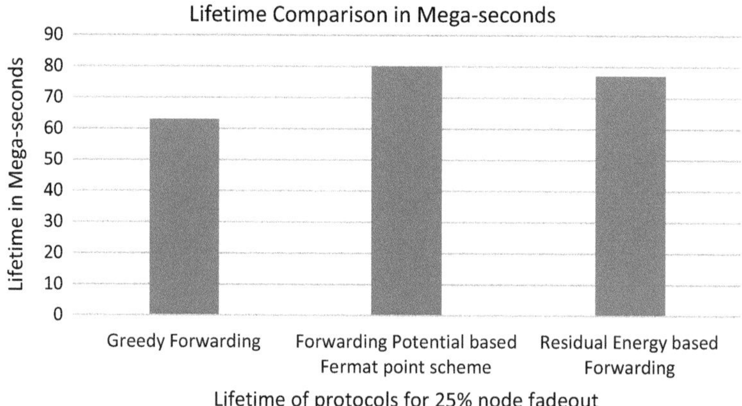


**Fig. 8.5** Lifetime comparison of a forwarding potential-based Fermat point scheme with greedy and residual energy-based forwarding

away. Here also, we can see that the scheme that uses the principles of Fermat point performed better as compared with the remaining two schemes.

For the results of Fig. 8.6, the n out of n definition has been considered. According to it, a network will only be considered alive if all the nodes are functional on the basis of their residual energy.

For this figure, however, all the considered protocols are Fermat point-based. It compares the KPS protocol with Fermat point-based variants of the following protocols: (i) Residual Energy-Based Forwarding (F-Residual), (ii) Compass Routing (F-Compass) and (iii) Greedy Forwarding (F-Greedy).

The result shows that KPS protocol proposed by Ghosh et al. (2016) edged over its Fermat point-based counterparts; viz. F-Greedy, F-Residual and F-Compass;

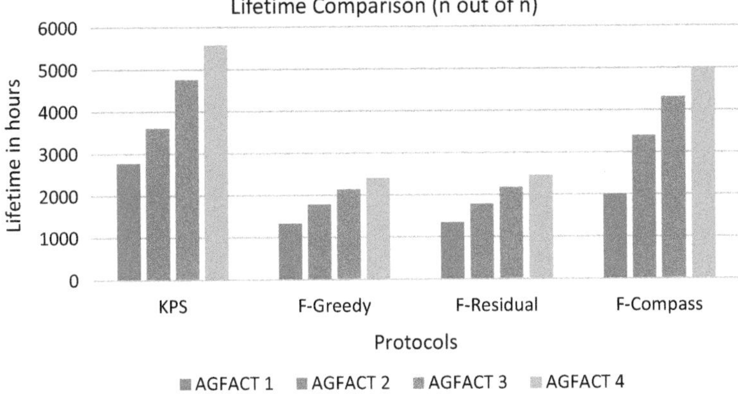


**Fig. 8.6** Lifetime comparison for different values of AGFACT

for different AGFACT values. The transmission range (TXR) of each node was considered as 80 m.

## 8.5 Conclusion

The results in this chapter showed that different Fermat point-based schemes ensure more lifetime as compared with other types of forwarding strategies. Therefore, in agricultural land, the deployed sensor nodes for measuring farm parameters may be embedded with a Fermat point-based routing protocol to ensure maximum operating time for the nodes along with cost amortization of the deployed network.

## References


Dietrich I, Dressler F (2009) On the lifetime of wireless sensor networks. ACM Trans Sens Netw (TOSN) 5(1):1–39

Ghosh K, Das PK (2013) Effect of forwarding strategy on the life time of multi-hop multi-sink sensor networks. In: Proceedings of the third international conference on trends in information, telecommunication and computing. Springer, New York, NY, pp 55–64

Ghosh K, Roy S, Das PK (2009, December) An alternative approach to find the Fermat point of a polygonal geographic region for energy efficient geocast routing protocols: global minima scheme. In: 2009 first international conference on networks & communications. IEEE, pp 332–337

Ghosh K, Das PK, Neogy S (2015) Effect of source selection, deployment pattern, and data forwarding technique on the lifetime of data aggregating multi-sink wireless sensor network. In: Applied computation and security systems. Springer, pp 137–152.

Ghosh K, Das PK, Neogy S (2016) Kps: a Fermat point based energy efficient data aggregating routing protocol for multi-sink wireless sensor networks. In: Advanced computing and systems for security. Springer, New Delhi, pp 203–221

Ghosh K, Neogy S, Das PK, Mehta M (2017) On regulating lifetime of a 3-sink wireless sensor network deployed for precision agriculture. Int J Next-Gener

Heinzelman WB, Chandrakasan AP, Balakrishnan H (2002) An application-specific protocol architecture for wireless microsensor networks. IEEE Trans Wireless Commun 1(4):660–670

Hwang IS, Pang WH (2007) Energy efficient clustering technique for multicast routing protocol in wireless adhoc networks. IJCSNS 7(8):74–81

Kassim MRM, Mat I, Harun AN (2014, July) Wireless Sensor Network in precision agriculture application. In: 2014 international conference on computer, information and telecommunication systems (CITS). IEEE, pp 1–5

Lee SH, Ko YB (2006, May) Geometry-driven scheme for geocast routing in mobile ad hoc networks. In: 2006 IEEE 63rd vehicular technology conference, vol 2. IEEE, pp 638–642

Madan R, Cui S, Lall S, Goldsmith A (2006) Cross-layer design for lifetime maximization in interference-limited wireless sensor networks. IEEE Trans Wireless Commun 5(11):3142–3152

Mafuta M, Zennaro M, Bagula A, Ault G, Gombachika H, Chadza T (2013) Successful deployment of a wireless sensor network for precision agriculture in Malawi. Int J Distrib Sens Net-Work 9(5):150703

Mekonnen Y, Namuduri S, Burton L, Sarwat A, Bhansali S (2019) Machine learning techniques in wireless sensor network based precision agriculture. J Electrochem Soc 167(3):037522

Sahitya G, Balaji N, Naidu CD (2016, July) Wireless sensor network for smart agriculture. In: 2016 2nd international conference on applied and theoretical computing and communication technology (iCATccT). IEEE, pp 488–493

Sahota H, Kumar R, Kamal A (2011) A wireless sensor network for precision agriculture and its performance. Wirel Commun Mob Comput 11(12):1628–1645

Ssu K-F, Yang C-H, Chou C-H, Yang A-K (2009) Improving routing distance for geographic multicast with fermat points in mobile ad hoc networks. Comput Netw 53(15):2663–2673

Wang N, Zhang N, Wang M (2006) Wireless sensors in agriculture and food industry—recent development and future perspective. Comput Electron Agric 50(1):1–14

Wark T, Corke P, Sikka P, Klingbeil L, Guo Y, Crossman C, Valencia P, Swain D, BishopHurley G (2007) Transforming agriculture through pervasive wireless sensor networks. IEEE Pervasive Comput 6:2

# Chapter 9
# Application of IoT-Enabled 5G Network in the Agricultural Sector

**Kaushal Mukherjee, Subhadeep Mukhopadhyay, Sahadev Roy, and Arindam Biswas**

**Abstract** The high speed 5G network may play a key role in the agriculture sector over the next century, helping to improve crop quality by applying the least man force. Farmers can be more profitable by using intelligent and reliable precession cultivation. The introduction ultra-high-speed communication networks into the farming sector will significantly improve agricultural employment. The 5G platform's artificial intelligence-based data hub offers precision agriculture strategies that are both versatile and productive. It will also enable for the autonomous implementation of different robotic farm machinery during the ploughing, plantation, as well as strategic planning aspects of livestock production, resulting in safe, trustworthy, eco-sustainable, and even the creation of automated farms. This chapter discusses about the importance of 5G networks in the agricultural sector and the advantages of 5G communication over the commonly used networks. The appropriateness of 5G networks in the farming sector, requirement of high-speed communication system and application of 5G; advantages of 5G; implementations of 5G in smart farming like live tracking, remote guidance, threat detection, statistics analysis, web directories; and upcoming opportunities has been discussed in this chapter.

**Keywords** 5G Network · Agriculture · Antenna · Cyber-Physical System · IoT Underwater

## 9.1 Introduction

Agriculture is most countries' main basis of income and takes a major part in their livelihood. Cultivation encompasses crop production, domestic animal breeding and farmland cultivation in order to provide essential commodities to the people (Derpsch et al. 2010). Agriculture is practised in various ways in various parts of the world,

K. Mukherjee (✉) · S. Mukhopadhyay · S. Roy
Department of Electronics and Communication Engineering, National Institute of Technology, Jote, Arunachal Pradesh 791113 , India

A. Biswas
School of Mines and Metallurgy, Kazi Nazrul University, Asansol 713340, India

 
A. Choudhury et al. (eds.), *Smart Agriculture Automation using Advanced Technologies*, Transactions on Computer Systems and Networks,
https://doi.org/10.1007/978-981-16-6124-2_9

with the primary goal of delivering safe and healthy food to the inhabitants. It virtually guarantees a country's food security while also providing raw materials for automotive industry (Gonzalez 2004). Modern farming, also known as modern agriculture, began in the eighteenth century that was popularly known as "The British Agricultural Revolution", was a period in which many changes to farming methods were taken in a short period of time, resulting in a significant increase in the crop production (Trautmann et al. 1985). Between 1900 and the 1930s, the first agricultural revolution occurred, with automated farming permitting every agriculturalist to cultivate adequate food for the inhabitants. This agricultural revolution saw the introduction of new farming approaches and methods such as soil supervision. In the 1990s, Green Revolution occurred after several decades (Melillo 2012). With the help of scientific advances, pesticides that require rather less water were introduced, permitting both agriculturalists to feed 155 individuals. The use of mechanized tools in agricultural practises was encouraged during the second revolution, which increased both the rates well as the amount of crop production (Saigo 1999). After the end of second revolution, another revolution takes place, likewise known as the green revolution, was the period while almost everyone started to use genetically modified crops, resulting in increased produce output (Byerlee 1992). As the world's population is increasing continuously, the rate of crop production must be increased quickly around the entire world. A massive demand must be met in an economical fashion deprived of worsening assets like water and electricity (Lipton et al. 1990).

Various types of sensors such as temperature sensor, soil pH testing sensor, humidity sensor, drones, pest detection sensor, soil fertility detecting sensor, etc., are connected to the internet and functioning independently or semi-automatically to perform specific tasks and gather data to better understand efficiency and predictability, which are included in the IoT technology. Smart farming employs artificially intelligent (AI) approaches to determine soil conditions and identify seeds for plantation depends on the nature of farmland, environmental conditions, availability of water, and other factors; to diagnose infestations in an initial phase and choose and implement the necessary pesticides to boost production. AI algorithms use computers to read and process information obtained from sensors from flying drones above farmlands in order to determine planted problems. It is really feasible to perform cultivation simpler with the assistance of such innovations. The 5G data connection offers a quick way (faster than the conventional available communication networks such as 3G, 4G etc.) of transmitting information by avoiding congestion, degradation, and interruption in the transmission process. Despite the fact that the 4G network provides fast speeds and communication, it would not be feasible to interlink every one of the gadgets used throughout intelligent forming in remote places at minimal cost of maintenance. The adoption of the 5G cellular networks and very high-speed communications has simplified the job of efficiently sharing data.

Section 9.2 describes the importance of IoT in the agricultural sector, Sect. 9.3 discusses about the utilization of 5G network in the agricultural sector, Sect. 9.4 analyses the data obtained by various wireless devices, Sect. 9.5 describes some artificial intelligence powered robots and Sect. 9.6 concludes the chapter.

## 9.2 IoT in Smart Agriculture

Cultivators and farming machineries are moving to the Internet of Things for enhanced information processing and analyses. The Internet of Things includes a variety of detectors those are programmed to execute a variety of tasks that propel the IoT forward. These detectors are placed throughout the field, enabling farmers to access specific information about soil, plants, climate, livestock, and other measurements through an advance handset, which is internet connected (Ayaz et al. 2019). Farmers now have access to data thanks to Internet of Things (IoT) technology on various aspects of smart farming, including soil mapping, need of irrigation, use of fertilizers, disease monitoring, harvesting process, yields estimation as well as controlling all things that need to be considered. Agricultural drones and land-based detectors are used to create topographic plots of farming areas and help in collecting relevant data obtainable in the area.

### *9.2.1 Advantages of Smart Farming System*

Farmers can also use this method to conduct a systematic study to determine the soil conditions like temperature, alkalinity, nutrient content, soil fertility, etc. All of these parameters can be correlated with historical information, as well as the kinds of cropping done previously, fertilizers and pesticides used, water levels, and so on (Sena et al. 2002). The smart farming system also contains indicators that can precisely observe soil characteristics, acidity, water content, and other factors, using which numerous tools and techniques can take measures to prevent soil erosion, acidification, toxicity, and other issues. Plots of the farming land are generated with the help of smart devices, which contains information about the soil of the scanned area, and are used to cultivate various types of crops that are cultivated on some specific type of soil.

### *9.2.2 IoT-Based Agricultural System*

IoT-based irrigation is regarded as the most effective irrigation factor system caused by water shortages. Water conservation systems based on the Internet of Things allow for a specific flow to fields. With the intention of measuring pressure in addition to mitigate the impact of various factors that influence crop temperatures and pressure, the crop water stress index was created (Jackson 1982). To measure the crop water stress index (CWSI) and acquire the site-specific irrigation index, various cognitive information such as temperature, satellite photos, thermal lack of appreciation, and so on are merged. VRI (variable-rate irrigation) addresses such a need to optimize irrigation schedules using sensory information.

IoT stable system is being utilized to measure biochemical configuration of the soil, like phosphorus, nitrogen, and potassium levels should be adjusted as required to make sure plant productiveness. Data collected, like nutrient concentration and weather, assist in determining the specific chemical fertilizer needed for crop growth, since superfluous amounts disturb the soil's fruitfulness. To identify pest infestations, devices such as robots, aircraft, and detectors are used, enabling for specific pesticide application. Estimation and real-time production tracking are easily possible by using 5G network. The overall cost and environmental considerations are greatly reduced when these IoT-based tools are used. A number of mobile applications can use every single detail from these connected systems and provide information about agricultural yield, harvest efficiency etc., and all these information are saved on a dedicated server.

## 9.3 The Use of 5G in Agriculture

A faster and more reliable internet connection is needed for agricultural IoT systems to operate. The new generation of mobile networks is struggling due to the weak coverage in rural areas, and even where high-speed connectivity is available, system fails leading to huge competition. Though in the United Kingdom, nearly 80% of rural communities were outside the 4G range, according to a recent survey (Palmer and Cooper 2013). The area covered by recent available generations of communication system in remote locations is inadequate in many countries. Furthermore, in certain developing nations, several farms operate through a vast number of connected smart devices and computers, which need a continuous, stable ultra-fast network link to transmit and receive a massive volume of information, and current mobile network architectures cannot meet these requirements. Several innovative technologies, such as huge MIMO, communication protocols, and smaller modules, are required to achieve these goals as well as provide efficient communications over longer distances. As a result, by allowing wider coverage, lower transmission time, lower input power, and inexpensive smart phones, the 5G communication system has been perfectly equipped to promote advanced cultivation (Chettri and Bera 2019). Precision farming is predicted to become more popular as technology advances. Agricultural sector farmers will benefit greatly as 5G penetration increases, allowing them to handle farms, cattle, and other assets from homes, due to their huge size, higher data rate, and reduced power. 5G technologies would help take IoT sensor networking to the next stage, paving the way for ground-breaking smart farming element development, as shown in Fig. 9.1. The key areas in the agriculture industry where the installation of a 5G communication system would be advantageous. In this section, we'll look at how 5G can be used in the agricultural sector.

**Fig. 9.1** Utilization of drones and artificial intelligence-equipped sensors in agriculture

### *9.3.1 Advantages of 5G Network Over Other Conventional Networks*

In farming systems, 4G/3G/ or other IoT wireless communication infrastructure is often used to communicate Internet of Things connected devices for information exchange, accurate evaluation, and precise estimation, among other things. Whereas the cellular technology connectivity framework showed promising results, there are certain constraints that may prevent the innovation from reaching its full potential in the farming sector (Stewart 1981). One of the most major limitations is the operational location. Remote areas and portions of a region containing multiple buildings really aren't covered by existing cellular connections. The existing communication networks such as 3G, 4G, 5G NR all operates in the 1–6 GHz of band, whereas the Millimetre Wave (MMW) Bands operate in the 26–70 GHz of frequency band. Due to the congestion of the lower frequency bands with the existing networks, real-time video streaming cannot be conducted with high-quality video from remote places, which, in turn, may result in wrong solutions to problems like spray of pesticides (Farooq et al. 2019). As the automated system required being control from a remote place through commands, this requires ultra-high-speed networks that can be obtained from the 5G MMW frequency bands.

### 9.3.2 *Application of Drones in the Agricultural Sector Based on 5G Networks*

Farmland specialists may respond to agriculturalists' concerns via remote service using high-definition videos, eliminating any necessity of a physical intervention and in comfort of their own home environment. Agricultural specialists may explicitly grab a live broadcast through real-time surveillance systems to gain accurate insights into the problem and offer needed assistance to cultivators for the improvement of farming techniques when and where required (Charania and Li 2020). Agricultural consultancy, environmental monitoring, fertilizer application, pesticide application, are among several of the facilities offered by consultation amenities. Utilizing several detectors installed onto multiple machineries and tools to calculate a wide quantity of factors in real time, 5G would introduce a new maintenance model known as smart precision agriculture. It will significantly minimize human error caused by defective hardware or system failure.

Drones are capable of capturing images with multiple spectrums, thermal images, infrared images, hyper-spectral images, and also able to scan a large area for 3D mapping with the help of sensors that work on liar technology. Drones are useful for conducting fast and effective livestock tracking and auditing. Various drones may be used to conduct different tasks over many acres of land. The drones that are used in the agricultural sector are mostly operated with lower altitudes (below 120 m) and work on a particular wireless link; as a result, they can able to cover an area of 3–7 km through a bulky transmitting antenna. On the other hand, the application of 4G communication network connection for radio control tends to alleviate the distance problem, enabling a drone to travel several kilometres away from its transmitter. Handling and locating drones have become much more difficult and time-consuming as the number of people using drones has grown. The 5G communication network can be integrated along with the monitoring of drone traffic to improve drone protection and safety by facilitating low-altitude drone connectivity. The 5G network will provide universal mobile broadband service on the land and in the clouds. The drones can fly over the farming lands by the farmers and can potentially be operated from anywhere in the world. Farmers can access actual information from drones, such as high-definition video feed. Drones do not need to hold a lot of computing capacity, because all of the data can be transmitted to the cloud and processed more quickly thanks to 5G technologies. Multiple drones can coordinate with one another to establish coordinated automated movement above a geographic location as well as to implement various operations efficiently with nominal loss of information. For judgement purposes, a vast volume of collected data may be processed and evaluated. A vast number of 5G-enabled automated drones will travel across the clouds to bring farm goods to one's doorstep thanks to the 5G wireless network's extensive network coverage area and secure connectivity.

### 9.3.3 5G Network-Based Augmented Reality and Virtual Reality in the Agricultural Sector

Farmers will be benefitted from the smart technologies in a variety of ways with the help of ultra-high-speed communication network like 5G. Via artificial intelligence-based 5G-enabled lenses along with advanced devices, augmented reality will provide assistive information like seed condition, livestock, as well as apparatus statistics, live camera stream of agricultural land with weather forecasts, condition of soil, aquatic environments, and more. The agriculturalist, be able to achieve valuable details like are the yields are contaminated or not, at what time the crops can be harvested by wearing artificial intelligence-based glasses and looking at the crops (Ali and Nahar 2018), as shown in Fig. 9.2. As a result, farmers can farm in a more sustainable manner, potentially reducing labour, time consumption, and crop losses while maintaining a high-quality yield. As a result, agriculturalists will perform cultivation in a much more efficient manner, which might reduce costs, manpower, resources like time as well as crop damages, which also safeguard a superior harvest.

The wider bandwidth requirements for online processing of agricultural land data are met by 5G networks. Everything could be accessed live through cloud with the help of ultra-fast 5G networks, eliminating unnecessary traffic. 5G cloud offers an

**Fig. 9.2** Artificial intelligence-based smart software application

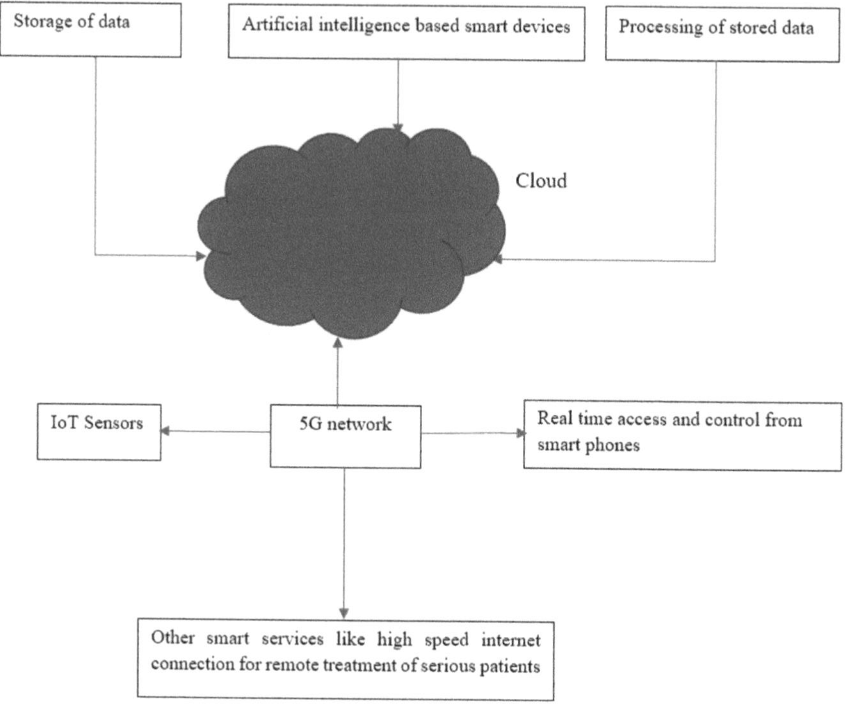


**Fig. 9.3** Real-time processing of on-field data through cloud

opportunity for highly interactive virtual reality services for agricultural education learning, including live video experiences, as shown in Fig. 9.3.

## 9.4 Data Analysis and Discussion

The implementation of IoT with 5G networks creates major progresses in real-time video tracking, remote analysis and on-site solutions to provision advanced agriculture, and it is also used to steady drones and automated machines by monitoring their parameters precisely. Automation in agricultural sector is rapidly improving, which, in turn, results in advanced solutions for the increment of crop production, manufacturing process whereas, among other things, artificial intelligence-powered automated machines are perfectly ready to convert the agricultural industry to a smart industry. Weeds are detected and eliminated accurately over a specific area by robots equipped with machine learning and artificial intelligence technologies, avoiding harming the crops. Robots are trained to use GPS to navigate around the field and locate and collect ready-to-harvest fruits and vegetables.

Computer vision technique is sometimes a fundamental ability of such automation, enabling the automated machines to see, classify, optimize, and end up taking a few smart system needed for distinct plants (Lu et al. 2020).

For navigation, such automated devices have a laser rangefinder for the detection of distance between two locations, along with a camera used for pattern recognition, in order to avoid collision.

Table 9.1 compares the performance of the network connectivity necessities of several virtual reality facilities. With a quick and secure mobile network, broadcast of input data and output of audio data back to the user can be done with ultra-low latency, giving users a real-time immersive experience without nausea. Once the automated device discovers pests by performing a thorough examination of certain yields, for example, it marks the positions and descriptions of the pests discovered. The automated devices are installed to monitor crop situation, navigate autonomously to a designated spot, and apply the appropriate insecticide depending upon the pests present. Face detection artificial intelligence-enabled automated robots have been used to recognize facial reactions of farm animals' wellbeing and to ring alarm bells when any anomalies are found (Johnson et al. 2014). It is often used to identify distinct animals so that their activities can be easily monitored and tracked. Most robots are capable of packaging harvested goods for shipping.

Because it is a supportive atmosphere with a closed structure, the greenhouse is a perfect location for several types of robots to thrive. In a greenhouse, a significant number of robotic arms may be used to perform several activities like planting, positioning, irrigation, and harvesting at a minimum amount of time.

## 9.5 Artificial Intelligence-Powered Robots

The automated devices can navigate their surroundings and be operated from anywhere using cloud computing and a vast volume of data transmitted over 5G. All of these robots will relay real-time images and videos obtained from various integrated detectors over a 5G mobile network with extremely low latency. Information is among the most critical aspects of the technological advancements that are propelling the intelligent farming sector forward. On several crops, most of the information gathered from various sources, like IoT cameras, aircraft, and robots, is stored in the local file system.

Data are among the most crucial elements of the technological advancements that are propelling the modern farming sector forward. On so many farms, most of the data obtained from numerous sources, like IoT cameras, drones, and robots, are stored in the cloud. 5G and artificial intelligence would allow for quick data transfers to the cloud, allowing for live analysis and interaction directly through the computers to help modernize and automatism the farming process as shown in Fig. 9.4 (Gill et al. 2019).

Such bigger data sets (ranging in size from Mbps to Terabytes) are always moved via various sources to the cloud and then returned to users including farmers and data

**Table 9.1** Comparison of the performances of 5G with existing technologies

| Network | Frequency range | Area covered | Speed | Bandwidth | Modulation technique | Latency | Input power | Mobility |
|---|---|---|---|---|---|---|---|---|
| Bluetooth | 2.40–2.48 GHz | 100 m | 3 Mbps | 15 MHz | GPSK/DQPSK/DPSK | Low | Low | No |
| Wi-Fi | 02.4–06.0 GHz | 100 m | 10 Mbps | 40 MHz | BPSK/QPSK/QAM | Low | Low | No |
| 2G | 850 MHz | Several km | 64 Kbps | 25 MHz | TDMA/CDMA | Low | Low | 60 km |
| 3G | 800–2100 MHz | Several km | 8 Mbps | 25 MHz | CDMA | Low | Low | 100 km |
| 4G | 1.7–2.7 GHz | 12 km | 300 Mbps | 150 MHz | CDMA/OFDMA | Low | Low | 200 km |
| NB IoT | 700–900 MHz | 15 km | 200 Kbps | 200 kHz | BPSK/QPSK | Low | Low | Yes |
| 5G NR | 01.0–06.0 GHz | 200 m | 1 Gbps | 400 MHz | OFDMA | Low | Low | 500 |
| 5G MMW | 26–70 GHz | 100 m | 20 Gbps | 4.2 GHz | OFDMA | Lowest | Lowest | 500 |

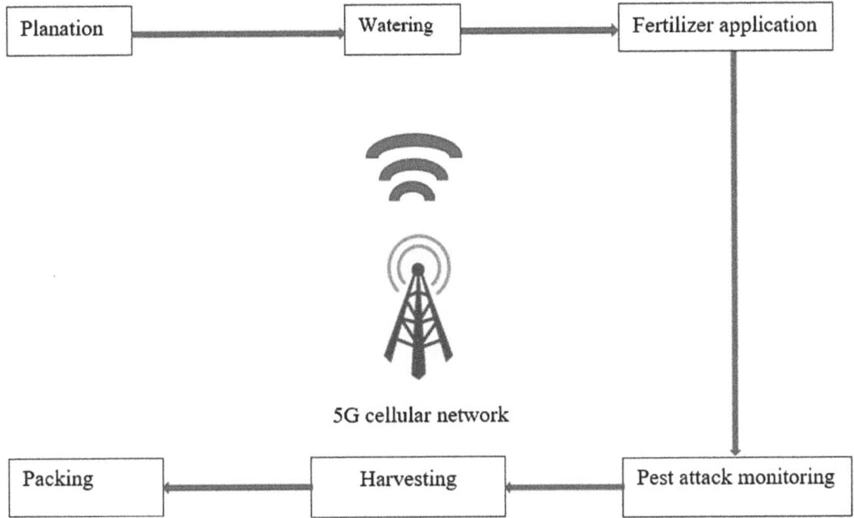


**Fig. 9.4** Application of smart robots in several fields of agriculture

analysts. For instance, after closely inspecting a crop and suspecting pests, the robots take a photograph of the infected crop and upload it to the cloud for pest detection (Wallace et al. 2018). The farmer receives the data, and the robots that are built for controlling the pest attack (Roy et al. 2021), and takes the necessary action.

In smart farming, the data are processed with the help of cloud in order to reduce the complexity of the system. The cloud acts as a data storing centre, which stores the information about the navigation data of the robots as well as the control commands for the robots. Agriculture intelligence (XAI) analyses these data in real time to create AI prescription maps for plant defence drones or automated agriculture machines using variable-rate application (VRA). Edge computing eliminates the need for graphics processing. Only 5G can provide such a higher data rate due to wider bandwidth needed to process the data (120 Mbps).

Application of 5G significantly reduced the size of the robots, complexity in the robot design, power requirement and manufacturing cost. 5G network would greatly increase data transfer speeds across the existing generation of communication networks. 5G offers a data transmission rate of 1–20 Gbps (Mukherjee et al. 2020) and with the help of these high-speed networks, a significant volume of data can be efficiently transmitted through several devices while minimising data loss, eliminating link downtime.

All of these data can be accessed safely in real time due to its cutting edge low latency. Cloud computing takes full advantage of 5G technology, allowing for quicker data processing in the cloud and low round-trip latency between different 5G connected devices, resulting in maximum smart farm efficiency. Sensors, drones, robots, mobile vehicles, cloud computing, and data analytics are all examples of 5G network applications in smart farming as shown in Fig. 9.5. All of these scenarios

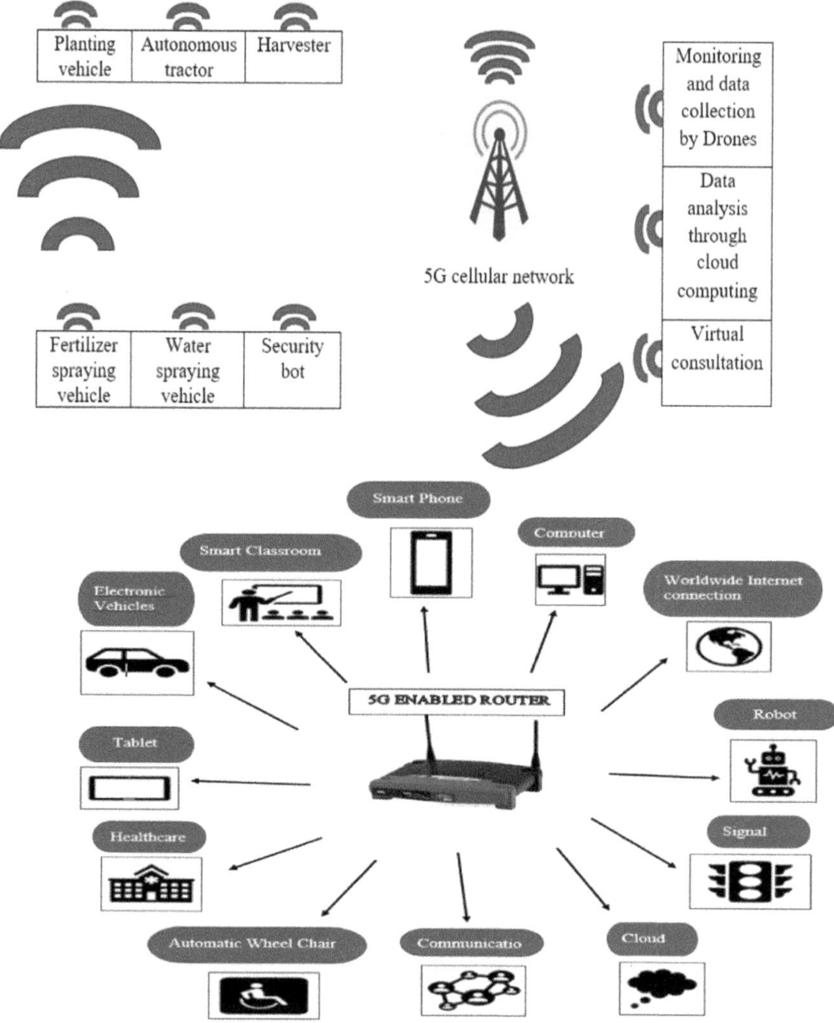


**Fig. 9.5** Utilization of 5G network in various fields of agriculture and other IoT-enabled devices

make use of the four main 5G characteristics: system size, extreme low latency, ultra-reliability, and stability. Drones, IoT sensors, and robots are only a few of the smart agriculture components that work together to increase efficiency and dramatically reduce costs.

## 9.6 Conclusion

Technology now plays a critical role in all aspects of society's growth, including building construction, vehicle manufacturing, aerospace, telecommunications, and military operations. In this regard, conventional fields like agriculture technology (here smart farming) generate higher crop yields with less human interference in a short amount of time. The introduction of 5G meets current requirements and demands in smart farming, allowing for increased production with less human labour. In all areas, all countries will adopt 5G networks; as a result, internet costs will be drastically reduced, and accessibility will be greatly improved. The use of 5G would greatly reduce the investment costs for smart farming, which will be a blessing to farmers. Cultivators would be perfectly ready for automated cultivation, by using their cell phones to anticipate and avoid crop disease. By expanding their physical infrastructure, mobile operators will make major assistances to smart farming. Sensors can capture data in the field, which will be stored in the cloud and analysed as required. Farmers will be able to increase their yields by several times thanks to automated technologies, artificial intelligence, advanced machineries, and cloud-based mobile applications. In the forthcoming years, connected farms would be built on the concept of linking, storing, and analysing vast volumes of data in order to boost productivity and performance. There is a good chance that the employment in the agriculture sector will decrease significantly. 5G network has not still explored all its features; innovative software and smart technology will be developed in the upcoming years.

**Acknowledgements** This work is a part of the Doctoral Research work of Mr. Kaushal Mukherjee, National Institute of Technology, Arunachal Pradesh, India, under the institute fellowship scheme. This research work is supported by Indian Institute of Technology Guwahati Technology, Innovation and Development Foundation, a Technology Innovation Hub (TIH) under NM-ICPS, Department of Science and Technology, DST/NMICPS/TIH12/IITG/2020/2(G).

## References

Ali M, Nahar L (2018) Automatic irrigation and monitoring system, Master Thesis, Daffodil International University

Ayaz M, Ammad-Uddin M, Sharif Z, Mansour A, Aggoune EH (2019) Internet-of-things (IoT)-based smart agriculture: toward making the fields talk. IEEE Access 7:129551–129583

Byerlee D (1992) Technical change, productivity, and sustainability in irrigated cropping systems of South Asia: emerging issues in the post-green revolution Era. J Int Dev 4(5):477–496

Charania I, Li X (2020) Smart farming: Agriculture's shift from a labor intensive to technology native industry. Internet of Things 9:100142

Chettri L, Bera R (2019) A comprehensive survey on internet of things (IoT) toward 5G wireless systems. IEEE Internet Things J 7(1):16–32

Derpsch R, Friedrich T, Kassam A, Li H (2010) Current status of adoption of no-till farming in the world and some of its main benefits. Int J Agric Biol Eng 3(1):1–25

Farooq MS, Riaz S, Abid A, Abid K, Naeem MA (2019) A survey on the Role of IoT in agriculture for the implementation of smart farming. IEEE Access 7:156237–156271

Gill SS, Tuli S, Xu M, Singh I, Singh KV, Lindsay D, Garraghan P (2019) Transformative effects of IoT, blockchain and artificial intelligence on cloud computing: evolution, vision, trends and open challenges. Internet of Things 8

Gonzalez CG (2004) Trade liberalization, food security, and the environment: the neoliberal threat to sustainable rural development. Transnat'l l Contemp Probs 14:419

Jackson D (1982) Canopy temperature and crop water stress. In: Advances in irrigation, vol 1. Elsevier, pp 43–85

Johnson DO, Cuijpers RH, Juola JF, Torta E, Simonov M, Frisiello A, Beck C (2014) Socially assistive robots: a comprehensive approach to extending independent living. Int J Soc Robot 6(2):195–211

Lipton D, Sachs J, Fischer S, Kornai J (1990) Creating a market economy in Eastern Europe: the case of Poland. Brook Pap Econ Act 1990(1):75–147

Lu Y, Xu X, Wang L (2020) Smart manufacturing process and system automation–a critical review of the standards and envisioned scenarios. J Manuf Syst 56:312–325

Melillo ED (2012) The first green revolution: debt peonage and the making of the nitrogen fertilizer trade, 1840–1930. Am Hist Rev 117(4):1028–1060

Mukherjee K, Das A, Roy S (2020) Comparative study on a U-slot miniaturized CPW-Fed multi-band antenna applicable for 5G communication. In: Electronic systems and intelligent computing springer. Singapore, pp 707–718

Palmer J, Cooper I (2013) United Kingdom housing energy fact file 2013. A report prepared under contract to DECC by Cambridge Architectural Research, Eclipse Research Consultants and Cambridge Energy. London, UK, Department of Energy & Climate Chan

Roy S, Mukherjee K, Biswas A (2021) Plane the IoT and machine learning, pp 249–269

Saigo H (1999) Agricultural biotechnology and the negotiation of the biosafety protocol. Geo Int'l Envtl l Rev 12:779

Sena MM, Frighetto RTS, Valarini PJ, Tokeshi H, Poppi RJ (2002) Discrimination of management effects on soil parameters by using principal component analysis: a multivariate analysis case study. Soil Tillage Res 67(2):171

Stewart RB (1981) Regulation, innovation, and administrative law: a conceptual framework. Calif l Rev 69:1256

Trautmann NM, Porter KS, Wagenet RJ (1985) Modern agriculture: its effects on the environment

Wallace RD, Bargeron CT, Moorhead DJ, La JH (2018) Forest,contractor's report

# Chapter 10
# An Economical Helping Hand for Farmers—Agricultural Drone

**Mainak Mandal, Ravish Jain, Aman Pandey, and Richa Pandey**

**Abstract** The world population is increasing day by day and is projected to reach 9 billion by 2050. Due to an increase in population, agricultural consumption will also increase. So Agri-drone is one of the most promising sectors, dealing with a lot of problems in which one of the main problems is manual spraying of pesticides. Pesticides sprayed manually have harmful side effects on human health. The common side effects are endocrine disruption, mild skin irritation to birth defects, tumors, blood and nerve disorders, genetic changes, etc. A better alternative to the manual spraying of pesticide is the use of drones mounted with a spraying mechanism. Basic structure of the machine would consist of: a motorized pump, 6 L of storage capacity tank, two nozzles for fine spraying, a Quadcopter configuration frame, suitable landing frame, four Brushless Direct Current (BLDC) motors with suitable propellers to produce required thrust (at 100% RPM) and suitable lithium–polymer (LI-PO) battery of current capacity 2500 mAh and 12 V to meet necessary current and voltage requirements. For monitoring the spraying process, a First-Person View (FPV) camera and transmitter are fixed in the drone and also for checking pest attacks on the plants. The drone will monitor along with a circular area and it will perform all operations as controlled by the operator. Image Processing algorithms are being used to find accuracy in monitoring the crop condition and controlling the spraying system to work accordingly. The purpose of this project is to provide a working model of the drone with a pesticide spraying mechanism that can not only reduce the risk involved in manual spraying of pesticides but also reduce the time, number of labor and cost of pesticide application.

M. Mandal (✉) · R. Jain · A. Pandey · R. Pandey
Mechanical Engineering Department, Birla Institute of Technology Mesra, Ranchi, Jharkhand 835215, India
e-mail: be10449.17@bitmesra.ac.in

A. Pandey
e-mail: be10629.17@bitmesra.ac.in

R. Pandey
e-mail: richapandey@bitmesra.ac.in



A. Choudhury et al. (eds.), *Smart Agriculture Automation using Advanced Technologies*, Transactions on Computer Systems and Networks,
https://doi.org/10.1007/978-981-16-6124-2_10

**Keywords** Agriculture · Unmanned aerial vehicle · Image processing · Fuzzy logic · Agri-drone

## 10.1 Introduction

Agriculture in India is the primary source of livelihood for about 58% of the population and is the backbone (Shivaji et al. 2016) of the Indian economy, so it is very important to improve the efficiency and productivity of agriculture along with safe cultivation of the farmers.

There are a variety of complex factors that can influence the growth of crops. These factors may include water access to changing climate, wind, soil quality, the presence of weeds and insects, variable growing seasons, and more. Agricultural drones are becoming modern tools to provide farmers with a lot of details about their crop conditions and to achieve some precise farming solutions for special types of crops and for limited areas and increase the overall profit. Drones can be used to collect the data related to yielding of crop, quality of soil, nourishment of crop and soil, weather-rainfall results and more.

Out of all the other areas in the agricultural sector where drones can play a very important role, pesticide spraying is one of the main areas that must be aimed at. The agricultural fields face huge loss due to diseases caused by pests and insects, which reduces the productivity of the crops. Pesticides and fertilizers are sprayed manually on the crops to kill the insects and pests in order to improve their productivity. This manual spraying of pesticides and fertilizers has a very serious health effect on the personnel involved in spraying. Therefore, in order to reduce the risk involved in manual spraying, an Unmanned Aerial Vehicle with some advancement can be made powerful, safe and efficient for spraying pesticides and can offer potential (Ahirwar et al. 2019) for facing several major or minor challenges. Image Processing and Fuzzy Logic controller (Baba 2015) to adjust the quantity of pesticide to be sprayed on the crops are proposed in this paper. The objective of our study is to develop a spraying system for an autonomous UAV that can be precisely used to spray the pesticides and also cover large areas of fields while spraying pesticides in a smaller interval (Shaw and Vimalkumar 2020) of time when compared with a manual sprayer. Thus, these advances in the field of agriculture can assist in improving crop growth and production and ease the life of the farmers.

## 10.2 Design and Working

The first step for designing this agricultural drone is to gather all the necessary parts of it, such as ESC, propeller, FPV camera, pump and the video transmitter. Then for stable design, we have to calculate the payload with respect to the weight of the payload motor. Then for operating the motor, a battery of appropriate current and

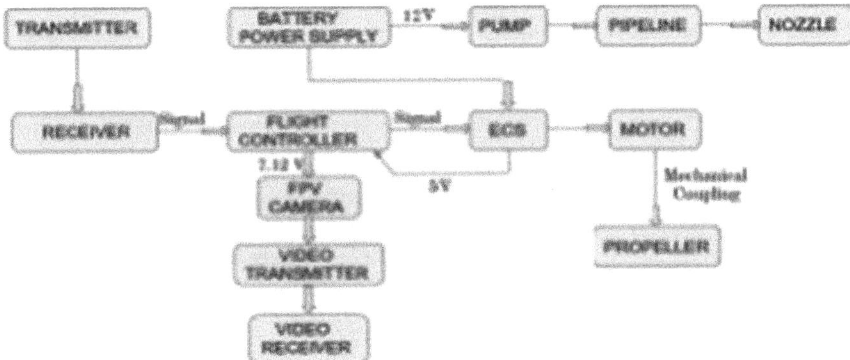


**Fig. 10.1** Block Diagram of working process

voltage is needed. Then the requirement of thrust force is calculated and the final frame of the drone is designed with four arms of required arm number and length. The weight of pesticide, weight of 6 L storage tank, pump and nozzles are estimated together as the payload.

Basically, the mainframe consists of four arms, four motors and four propellers. Each arm has a fixed motor at its free end, and the propellers are coupled to the motor mechanically. The output side of all the ESCs is connected with the motors and the input side of the ESC is connected with the flight controller. The power distribution panel where the power is supplied by Li-Po batteries is connected with other inputs of ESC. The signals from the transmitter are received by the receiver connected with the flight controller. Also a FPV camera is attached to the flight controller. An inclined base storage tank of $20 \times 20$ cm$^3$ dimension is mechanically coupled to the mainframe so that it can be drained completely when needed. The plastic tubes and nozzles are fixed between each other. In the spraying system, the pump is driven by the power distribution board. The inlet of the pump is connected with the storage tank, and the output of the pump is connected to the plastic tubes where nozzles are fixed.

The working process is as follows, the signal is transmitted from the transmitter and received by the receiver end. From there, the signal passes to the flight controller where the signal gets processed with accelerometer and gyroscope sensors. Then a modified signal is then sent to the ESC, where only the specific amount of current goes to the motor as it receives the signal. Mechanically coupled propellers then rotate to produce the required amount of thrust. With the help of the current supply from the flight controller, the FPV camera records video and the video signal gets processed by the transmitter–receiver zone. Current is supplied from the battery to the pump to pressurize the fluid in the storage tank and then it flows through the pipeline and enters the nozzle to get sprayed. Input current is used to vary the flow rate for the pump. Finally, the material selection for the body part is very crucial because it plays a very important role in determining the drone's stability and efficiency.

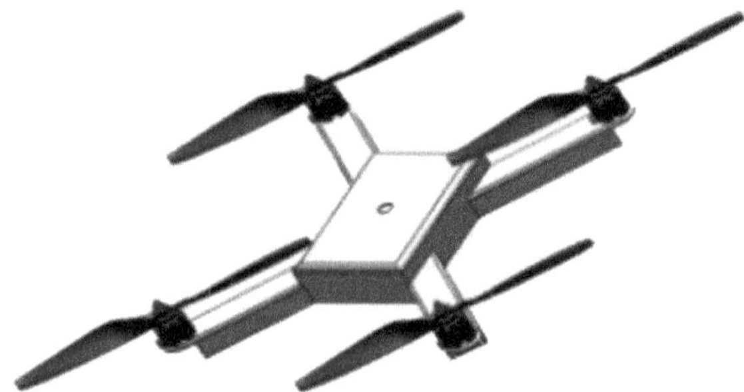

**Fig. 10.2** Basic Working design of a drone

This basic approach includes the manual analysis of crops and it is dependent on the farmer, which is a very rigorous work. Thus to reduce manual workload, this paper suggests a image processing-based approach, which uses Fuzzy Logic for better analysis of crops.

## 10.3 Fuzzy Logic

Diseases are bound to affect the crops and their treatments involve spraying of chemical pesticides (Mahajan and Dhumale). Manual analysis of crops is dependent on the farmer and is a tedious task. The amount of pesticides used needs to be measured as if not given in appropriate quantities can cause the crops to die or may not even cure the disease. Development in the fields of robotics, sensor networks and information technology helped push forward the concept of precision agriculture. It involves automation of agricultural processes thereby reducing cost of operations and adjusting the requirements according to the demand. This paper works on two diseases, Bacterial spot and Early Blight, that commonly occur in Tomato crops (Figs. 10.3 and 10.4).

There are numerous other diseases in Tomato crops with their grading scale (Manual for Tomato Pest Surveillance) for which the methodology being explained

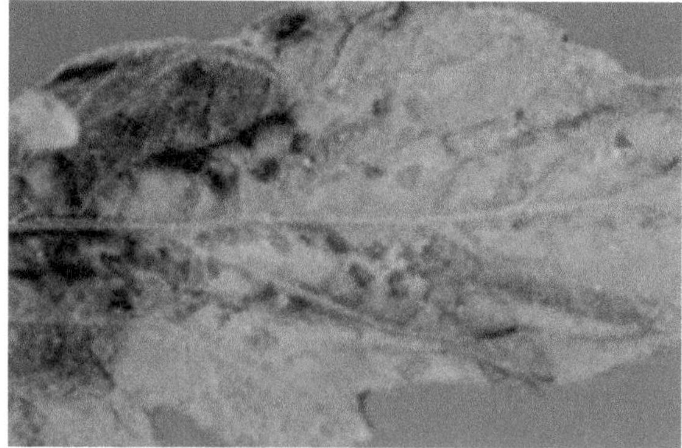

**Fig. 10.3** Early blight (*source* pestnet.org)

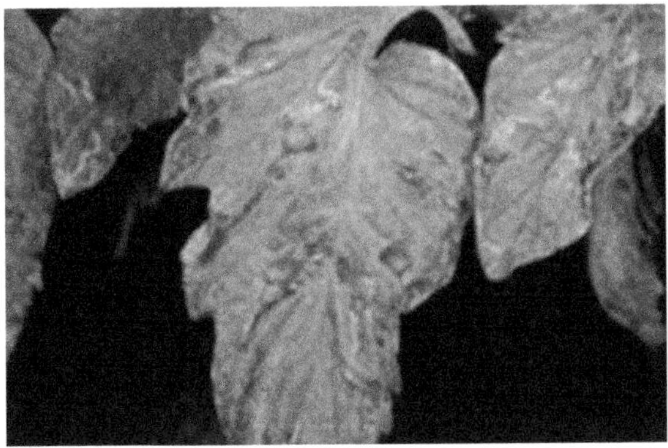

**Fig. 10.4** Bacterial Spot (*source* pestnet.org)

in the paper can be customized to meet the needs. Presently, the percent infection is calculated through manual surveillance and is prone to human errors.

Any normal modeling technique such as template matching, supervised or unsupervised learning would not produce satisfactory results in classification of agricultural products. This requires self-learning techniques such as neural networks or Fuzzy logic (Sabrol and Kumar).

## *10.3.1 Methodology*

For purpose of experimentation, tomato leaves are taken.

### 10.3.1.1 Image Pre-processing

After acquiring the images, they were resized and filtered to improve the resolution and remove the noise. **Gaussian filtering** or Median filtering among many other filtering techniques can be used to smooth the image and remove noise. Gaussian filter is a linear filter that blurs the edges and reduces contrast whereas Median filter is a non-linear filter.

### 10.3.1.2 Image Segmentation

It is the process of splitting the digital image into its constituent regions or objects so as to obtain suitable and easy to analyze. The objective of segmentation is to obtain the object of interest, which is the defective region on the leaf. Techniques such as compression methods, clustering-based methods, threshold methods, histogram-based methods, region growing methods, etc. are used for image segmentation. Otsu thresholding method and *K*-means clustering both prove to be excellent choices for segmentation, the former uses global thresholding while the latter uses local thresholding method. Paper uses the color-based *K*-means clustering method for separating the defective image and segmenting it into its constituent objects or colored parts.

a. Otsu's Thresholding: It's an automatic image thresholding technique that involves separating pixels of image into foreground and background through a single intensity threshold. It works best when the image histogram is of bimodal distribution and has a deep sharp valley between its two peaks.
b. *K*-means Clustering: The algorithm clusters "*n*" number of objects based on characteristic into *k* partitions, where $k < n$. In layman terms, it is an algorithm to classify or to group objects based on their corresponding features into *k* numbers of groups. Here k is always a positive integer number (Mahajan and Dhumale). In general, the groups are created based on the rule of minimizing the sum of distances between the centroid of clusters and data. Steps involved in color-based *K*-means image segmentation are:

    i. Reading the image.
    ii. Converting the image from rgb to L*a*b color space.
    iii. Classifying the colors in '*a*b' space using *K*-means clustering.
    iv. Create images that segment the infected area on the leaf by color.

    Highlight the cluster with infected areas from among k different clusters.

#### 10.3.1.3 Calculation of Total Leaf Area and Defected Area

Total area of an image is equal to the number of pixels on it. Similarly the area of the defective part is calculated. For this purpose, the images are converted into binary images to find out the pixels on the image that are on.

#### 10.3.1.4 Percentage Infected Area

The ratio of area of defective part to the total area gives us the percentage infected area. This is further utilized in forming a fuzzy inference system to make our spraying system work.

#### 10.3.1.5 Fuzzy Logic

Any real-life problem might not have discrete conditions and may contain nonlinear relations between input and output categories. It resembles human decision-making methodology. Rather than taking a statement to be completely true or false it considers the degree of truth or takes approximate reasoning into consideration.

It provides means of translating qualitative and imprecise information into qualitative (linguistic) terms (Otsu). Fuzzy set theory and fuzzy rules help to show and process human knowledge in the form of IF-then rules. It has found application in a large number of domains such as process control, pattern recognition and classification, decision-making and operation research. The ability of Fuzzy logic is to be used in places where definite mathematical solutions can't be used and helps in a wide range of precision agriculture areas. Thus, it has been used to grade disease.

### *10.3.2 Results*

The sample image after pre-processing involving resizing and Gaussian filtering was sent for image segmentation (Fig. 10.5).

In $K$-means clustering, k or the number of clusters needs to be selected. Here, we took $k = 3$. Following clusters were obtained as shown in Fig. 10.6 and the infected region in Fig. 10.7.

Now from the two images, the area of the green part ($A_G$) and the infected part ($A_I$) were calculated.

- %age infected area = $A_I/(A_I + A_G)$
- $A_I = 6.3223e + 03$; $A_G = 48{,}494$
- %age infected area = 11.53%

Figure 10.8 shows the rules of the FIS for an infected area of 11.5% grades the disease at 2.64.

Now based on Table 10.1, a fuzzy inference system is created such that it meets the grading conditions. This will further be used to meter the dispensing of herbicides.

**Fig. 10.5** Original image after resizing and Gaussian filtering

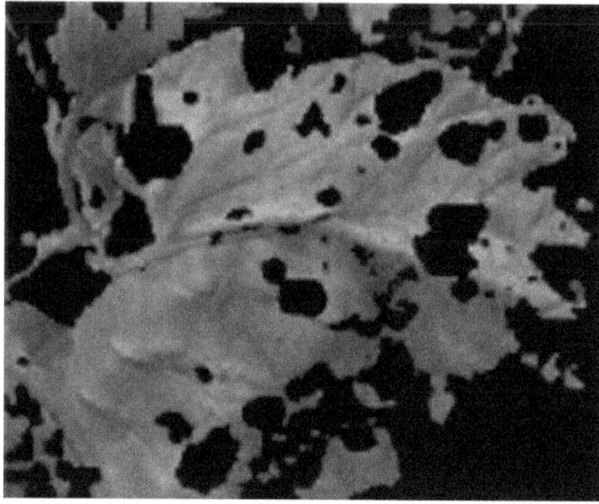

**Fig. 10.6** The green or uninfected region on the leaf

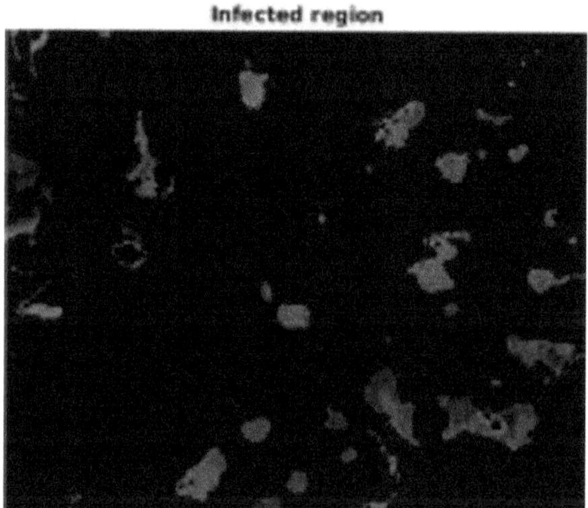


**Fig. 10.7** Infected region on the leaf

Automating the grading system gives fairly accurate results compared with the manual grading. The painstaking and time consuming manual grading system can be substituted by the use of image processing and fuzzy logic (A Comparative Study of Otsu Thresholding).

## 10.4 Spraying Mechanism

An efficient spraying mechanism (Fig. 10.9) should be used, which can help us to uniformly spray the pesticides on the crops.

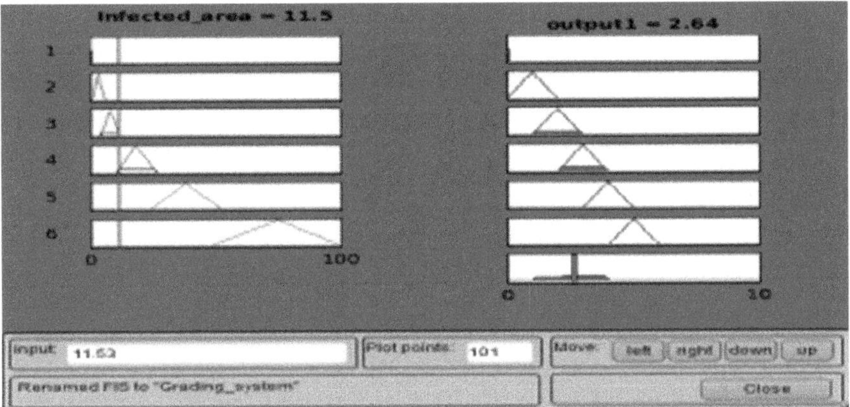


**Fig. 10.8** Grading System for calculating Infected Area

**Table 10.1** The grading scale for both the diseases is as follows (Manual for Tomato Pest Surveillance)

| Scale | Description (percentage area of plant infected) |
|---|---|
| 0 | No symptoms |
| 1 | 1–4% |
| 2 | 5–10% |
| 3 | 11–25% |
| 4 | 26–50% |
| 5 | > 50% |

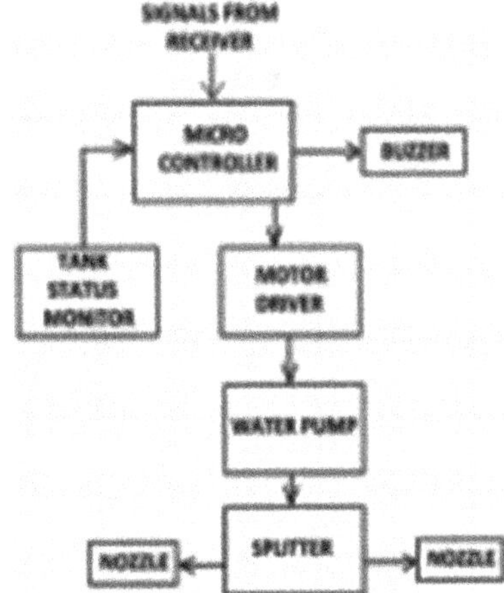


**Fig. 10.9** Block diagram of the spraying mechanism

(1) We can use piston crank mechanism in which the crank will be automated by motor due to which the piston will reciprocate and during its motion at its maximum displacement is strikes the spraying nozzle and using this mechanism we can make a spraying system spray the pesticide regularly after a fixed time interval.

(2) In this, the spraying mechanism has the following parts; of an Arduino UNO microcontroller—which is programmed to perform various functions, a tank-for the storage of pesticide, to which a water pump is connected, a splitter is connected to the water pump, which spits the pesticide to the two nozzles connected at the two opposite ends for spraying the pesticide. For controlling the speed of spraying, a motor driver circuit is used, and for detecting the level of pesticide in the tank, a pesticide level indicator circuit with buzzer is used.

The main functions of the spraying system are:

### *10.4.1 Pump ON/OFF Control*

The water pump that is used to spray the pesticide can be turned ON/OFF by sending control signals to the motor driver circuit from Arduino.

### *10.4.2 Spraying Speed Control*

The speed of spraying is achieved by sending a Pulse Width Modulation (PWM) signal to the motor driver IC.

### *10.4.3 Tank Status*

The amount of pesticide in the tank can be checked using a water level sensor. If the pesticide level reaches below a certain level, it can be notified to the operator by sending a control signal to the Arduino, which, in return, turns on the buzzer. Hence, when the buzzer is heard by the operator, he can land the drone for refilling.

## 10.5 Future Scope

1. Weight lifting capacity of the drone can be increased by (a) increasing the number of motors, (b) increasing the propeller size, (c) increasing the rpm of the motor.

2. Battery capacity can be improved to increase the flight time.
3. Increasing the size of the tank would lead to an increase in pesticide carrying capacity.
4. Arrangement of nozzles in the form of arrays could be used to cover large areas.
5. Angle of spraying can be controlled for accurate spraying.

## 10.6 Other Uses

Apart from agriculture drone, it can also be used in different areas:

1. In a pandemic situation such as the current COVID outbreak, spraying drones can be used for sanitization drives in hotspot areas.
2. Riot control drone.
3. Search and rescue operations can be conducted by thermal drones.
4. Disaster management systems can use UAVs for information gathering and supplying relief packages.
5. Express shipping and delivery.
6. Geo-Mapping of inaccessible location and terrains.

## 10.7 Conclusion

In this paper, we present our study on agricultural drone and its potential to improve the yield of crops by using some advanced techniques of Fuzzy Logic and Digital Image Processing in order to reduce the quantities of applied chemical materials over the green crops. This smart system may be considered as an economic and ecologic choice for the farmers. This advancement in spraying technique reduces the amount of pesticides being sprayed outside and leading to soil. It also improves the performance of the Agri-Drone by covering large areas and spraying the appropriate quantities of pesticide in a smaller span of time when compared with the classical spraying method. This drone can also be used in spraying disinfectant liquids over buildings, water bodies and highly populated areas.

## References

A comparative study of Otsu thresholding and K-means algorithm of image segmentation (2019)

Ahirwar S, Swarnkar R, Bhukya S, Namwade G (2019) Application of drone in agriculture. Int J Curr Microbiol Appl Sci 8(1), ISSN: 2319-7706

Arivazhagan IS, Shebiah RN, Ananthi S, Varthini SV (2013) Detection of unhealthy region of plant leaves and classification of plant leaf diseases using texture features. Agric Eng Int: CIGR J 15(1):211–217

Baba A (2015) Fuzzy logic based pesticide sprayer for smart agricultural drone. J Appl Sci

Basar S, Ali M, Ochoa-Ruiz G, Zareei M, Waheed A, Adnan A (2020) Unsupervised color image segmentation—a case of RGB histogram based K-means clustering initialization, Unsupervised color image segmentation: a case of RGB histogram based K-means clustering initialization

Image Thresholding (2017)—OpenCV-Python Tutorials 1 documentation

Image Processing Toolbox™ 7 User's Guide, ©Copyright (1993–2010) by TheMathWorks, Inc

Mahajan V, Dhumale NR(2018), Leaf disease detection using fuzzy logic, leaf disease detection using fuzzy logic

Manual for Tomato Pest Surveillance, National initiative on climate resilient agriculture (2011). TOMATO Prelims final.pmd

Otsu N (1979) A threshold selection method from gray-level histograms

Priyanka Shivaji C, Komal Tanaji J, Aishwarya Satish N, Mone PP (2016) Agriculture drone for spraying fertilizer and pesticides, 2(6), June-2017. Author F, Author S (2016) Title of a proceedings paper. In: Editor F, Editor S (eds) Conference 2016, LNCS, vol 9999, pp 1–13. Springer, Heidelberg

Pujari JD, Yakkundimath R, Byadgi AS (2014) Identification and classification of fungal disease affected on agriculture/horticulture crops using image processing techniques, In: IEEE international conference on computational intelligence and computing research, Coimbatore, pp 1–4

Sabrol H, Kumar S (2019), Plant leaf disease detection using adaptive neuro-fuzzy classification, (PDF) plant leaf disease detection using adaptive neuro-fuzzy classification

Shaw KK, Vimalkumar R (2020) Design and development of a drone for spraying pesticides, fertilizers and disinfectants. Int J Eng Res & Technol (IJERT) 9(5), ISSN: 2278-0181

Teknomo K (2007) K Mean Clustering Tutorial, Teknomo, Kardi. K-Means Clustering Tutorials. https://people.revoledu.com/kardi/tutorial/kMean/

Yang, CC, Prasher, SO, Landry, J-A, Perret J, Ramaswamy HS (2000) Recognition of weeds with image processing and their use with fuzzy logic for precision farming. Can Agric Eng 42(4)

# Chapter 11
# On Securing Smart Agriculture Systems: A Data Aggregation Security Perspective

**Tala Almashat, Ghada Alateeq, Arwa Al-Turki, Nora Alqahtani, and Anees Ara**

**Abstract** Due to the continuous growth in world population, there are significant issues upcoming related to food, health, and education. The most common problem is related to the production and supply of food from agricultural sector. With the recent advancements in technologies, the smart agricultural systems have evolved as an immediate solution for increasing the growth of food production and to meet the market demand. The smart agricultural systems are engineered by combination of sensors that coordinate and monitor the agricultural field's basic needs like humidity of soil, temperature of field, growth of the crop, etc. These sensors receive the commands from the user application and generate the sensed data to the gateway node and the results are analyzed by the field monitoring teams. Though there is an automated monitoring process, there is a high risk of data breach happening in these remote unstable environments. In this chapter, we aim to conduct a study on how to secure the smart agricultural systems by initially making a systematic risk assessment on the system. Based on the vulnerabilities and risk studied, we propose a light weight Arduino-based smart agriculture model. This model focuses on secure data aggregation using rc4 algorithm. Finally, security analysis is conducted on the proposed model and a project-specific checklist is summarized that highlights the recommendations for securing the smart agriculture systems.

**Keywords** Smart agriculture system · Cyber-physical systems · Risk assessment · Arduino-based light weight systems · Symmetric encryption

## 11.1 Introduction

The world population is increasing and there are more humans alive right now than ever before. With that number still growing, projections show that by 2050 the earth will have more than 9.7 billion human beings (Lyu et al. 2019). This magnificent

T. Almashat (✉) · G. Alateeq · A. Al-Turki · N. Alqahtani · A. Ara
Riyadh, Saudi Arabia

A. Ara
e-mail: aara@psu.edu.sa; 216410370@psu.edu.sa



A. Choudhury et al. (eds.), *Smart Agriculture Automation using Advanced Technologies*, Transactions on Computer Systems and Networks,
https://doi.org/10.1007/978-981-16-6124-2_12

population comes with a lot of challenges in production. With that fact, the food production must be improved using technology. Agriculture is one of the oldest industries ever and by introducing technology to the industry the production percentage will increase significantly. Smart farming consists of integrating advanced technologies into the basic farming practices to increase quality, speed, and efficiency. A major benefit of the smart system is that it reduces heavy labor and tasks for the workers. The smart agricultural system has great impacts on several stakeholders in the ecosystem such as farmers, consumers, agriculture cooperatives, food industries, government agencies, and nations critically dependent on agriculture.

The smart agricultural system consists of physical sensors that receive commands from user applications and generate data. The devices that are on the farm are connected to gateway supported edge nodes. These help in the communication between devices and help in filtering data and analytics. Additionally, the cloud holds a large amount of data such as monitoring information, energy management data, and information about the environment. All of these components work together to benefit the agricultural sector in various ways including sensing the moisture of the soil and the temperature of the environment to take specific actions. As a result of all the systems working together, security issues and risks arise that could affect each part of the system. This research paper aims to show how these components collect the data and the actions that are required based on the data. Additionally, it aims to show how to secure this system to protect the agricultural sector from unwanted attacks.

The chapter is organized as follows: Sect. 11.2 describes the risk assessment of smart agricultural systems, Sect. 11.3 proposes design and modeling of the secure smart agricultural system using Arduino boards, and Sect. 11.4 includes the security analysis of the proposed model. Finally, we summarize with the project-specific checklist that highlights the needs for securing the smart agriculture system.

## 11.2 Risk Assessment

### *11.2.1 Overview*

Risk assessment is a process to determine the probability of losses by analyzing potential hazards and evaluating existing conditions of vulnerability that could pose a threat or harm to property, people, livelihoods, and the environment on which they depend. In this section, we highlight the risks associated with agricultural cyber-physical systems. The components of the system were identified and documented, where we performed the risk analysis to identify where and how to secure the system itself (Fig. 11.1).

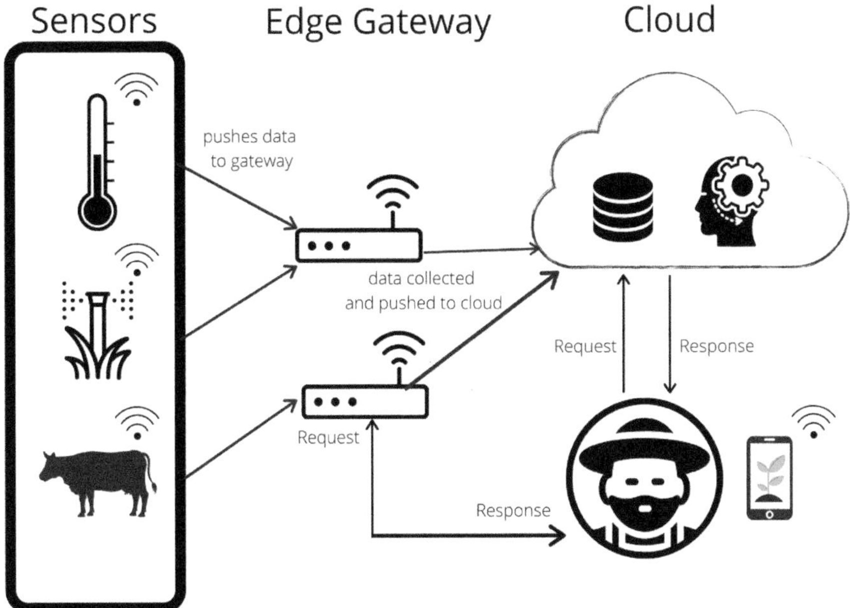


**Fig. 11.1** Shows an end-to-end interaction among various entities involved in the smart agricultural ecosystem

## 11.2.2 *Risk Assessment*

As discussed before risk assessment is a process of determining the probability of losses in any environment, more precisely here in smart agricultural system. Before getting deep into smart agriculture risk assessment, we want to understand the concept of risk assessment and the cycle of it. To start off, as we've studied before, risk assessment comes after the first step of risk management, which is identifying possible risks. Risks are measured by their potential severity of damage and loss and the possibility of their existence. This right here is what risk assessment is, the idea of measuring a risk and its effect on my assets. Although that is the basic definition of risk assessment, in cyber-physical systems we have a slightly different perspective of risk assessment.

As we know, cyber-physical systems are what we like to call a combination of cyber systems and physical systems. So, we cannot just apply the general risk assessment model on CPS systems. For CPS risk assessment, we have a detailed model we prefer to follow, which is broken down into three categories or specialties. Asset Identification, which answers the question "What might happen to the asset after the occurrence of the risk?" Threat Identification, which answers the question of what the possibility is, and lastly, the vulnerability identification, which answers the consequences question. These three aspects of risk assessment help identify the safeguards and the risk calculations. Table 11.1 shows the CIA trait priority level

Table 11.1 Security objectives priority to both systems

| Priority | CPS | IT systems |
|---|---|---|
| High | Availability | Confidentiality |
| Medium | Integrity | Integrity |
| Low | Confidentiality | Availability |

in CPS and IT systems, which are not similar. In which IT systems aim to have the confidentiality level high, and the availability level low, which in CPS look at it in the opposite way.

Besides these three aspects that create the risk assessment model of CPS, there are two characteristics researchers have looked into in CPS risk assessment, which is the safety and security of risk assessments in CPS.

#### 11.2.2.1 Risk Assessment Related to Safety

Safety risks come from several ways, the contact between the environment and CPS, and also the contact between authorized/registered users and the CPSs. In CPS, availability is our high priority security objective, in which safety comes along. These two objectives are proposed to preserve a stable state beneath the tolerable boundary.

#### 11.2.2.2 Risk Assessment Related to Security

In conjunction with the wide use of CPS applications, there is more data transmission happening on an hourly basis. This means that the security risk assessment importance leaps up in cyber-physical systems. Making that a reason that researchers have identified safety and security as the topmost important aspects of CPS risk assessment.

As in our topic, smart agriculture, we can apply the previous risk assessment methodologies. A couple of the risks that we may face in the smart agriculture environment would vary, from the production risks, marketing risks, to the human resource management risks. All of these risks can be evaluated and broken down based on their value and effect on the asset. Smart agriculture is one of the applications of CPS systems, which is the concept of controlling farms using modern ICT to increase quality and quantity of production. As we assess CPS systems using whichever methodology, we would assess smart agriculture in an equal way.

### 11.2.3 Assets Identification and Prioritization

In the smart agricultural environment, not all assets have the same criticality for stakeholders and service providers to do their function. An asset is marked as critical where any disruption will have a damaging effect on the activity of the overall system and ecosystems. As presented in Table 11.2, the properties were measured on the basis of the effect and severity of any disruption or assault to their operation.

Sensors are the most important physical assets in the smart agriculture framework. This can be demonstrated by the excellent role they have in the control of the smart agricultural climate. They collect the data needed to be shared, such as tracking information, energy management data, and environmental information. If any interference occurs to the sensors, the data cannot be transmitted correctly or at all, which is why the sensors are identified as the top priority. Sensors require a means through which the data is transmitted. So the second highest priority is the Base Shield. Sensors may also be installed in devices used by stakeholders, and these devices are connected to other devices used by ecosystem providers so that those devices take precedence over the network.

**Table 11.2** Smart agriculture simulator assets

| Asset | Functionality | Importance |
|---|---|---|
| *Hardware* | | |
| Base Shield | Use for connection of any microprocessor input and output pins. In this case it must be connected to the Arduino UNO to connect the sensors and receive the data | High |
| Arduino UNO | Consists of both a physical programmable circuit board and a piece of software, or IDE (Integrated Development Environment) that runs on the computer, used to write and upload computer code to the physical board | High |
| SparkFun soil moisture sensor | Measuring the moisture in soil and similar materials | High |
| DHT22 temperature sensor | To sense the temperature surrounding the agriculture | High |
| PIR sensor | Motion sensor. Used to detect whether a human or animal has moved in or out of the sensors range | High |
| Laptop | | High |
| *Software* | | |
| Data | The information gathered from sensors | High |
| Code | The code written in Arduino to control the sensors and receive information | |
| *Network* | | |
| Network | To connect the laptop to board (wired network). The protocol for the network is out of scope | Medium |

In the field of technology, CPS is well known for its ambiguity because it blends physical and cyber worlds together. As smart agricultural environment systems are CPS applications and contain the same complex components as other CPS applications such as smart grid, as time passes and technologies advance, they may become much more complex, meaning that such systems may have more valuable assets to be shielded from outsiders, or may get rid of some because of the technology sector changes.

As mentioned earlier, not all assets have the same criticality for the patients and facilities they require, but this will likely improve in the future as more and more assets will be implemented due to technical advances and CPS, so the asset reprioritization strategy will be needed in a manner that balances and adapts to new improvements in the wireless sensor network.

## *11.2.4 Malicious Actors and Threats Identification*

According to M. Gupta et al., smart sensor-based technologies and cloud supported applications in agriculture are being adopted more and more lately and with that, the dangers and possibilities of attackers orchestrating cyber-attacks on the system are rising. This section discusses the potential cyber-attacks in smart agriculture. Attacks are categorized in three different classes based on the CIA and STRIDE concepts (Gupta et al. 2020).

### 11.2.4.1 Data Attacks Which Affect the Confidentiality of the System

Data attacks include "*Insider Data Leakage*" which can lead to *Information Disclosure* that can occur from an inside employee that wants to harm the system. This information can be about the purchase of crops and soil information that could get in the hands of competitors of another country. Farmers fear this because the loss of confidential data can lead to the attacker using this data against them. Another example of the information that can be harmful if available to the attacker is anti-jamming devices information that helps the attacker in bypassing security measures. Also, an Attack against the cloud can cause "*Cloud Data Leakage*", because smart agriculture has sensitive and confidential information that could harm the country's farming system, attacks on the cloud could lead to serious economic losses.

### 11.2.4.2 False Data Injection Attacks Which Affect the Integrity of the System

Data *Tampering* attacks, where the attacker attempts to change some data of the system's critical data that could lead to the change of critical decisions. For example, changing the detected measurements of the moisture of the soil which could lead to watering the crops more which eventually damages the crops.

#### 11.2.4.3 Denial of Service Attacks Which Affect the Availability of the System

There are a large number of IoT devices and interconnected nodes in the agriculture systems and thus, they can be used by attackers to do a Denial of Service (DoS) attack. DoS attack can be used to stop the system's functionality and can cause serious damage to a large agricultural system of a country. DoS attacks in this case can disrupt a national economy (Fig. 11.2).

### *11.2.5 Vulnerability Identification and Quantization*

Vulnerability is considered a flaw in infrastructure, applications, firmware, operating systems, networks, entities, and processes that can be exploited. The latent likelihood of failure and harm of an agricultural vulnerability is commonly defined as Climate instability and the occurrence of an intense climate-related agricultural system. An occurrence and any entity or social group's susceptibility to its effects. Therefore, the definition of agricultural vulnerability focuses on the extent of physiological effect on crops induced by instability in the impact of different climatic elements on farmers.

Vulnerabilities have been identified using specialized scanning techniques. Many of these resources exist for testing and scanning the actual security status of the network, such as checking for open ports, unpatched applications, and any other

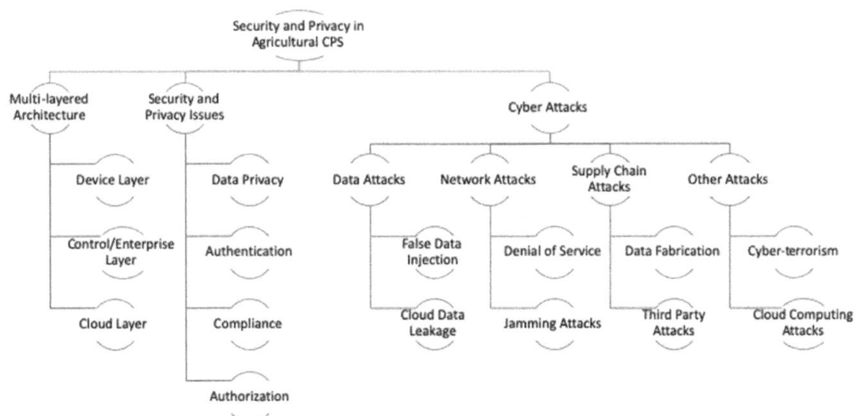


**Fig. 11.2** Most possible threats against smart agricultural system

bugs. A few of these programs concentrate on a particular network process or region, while others may search the whole network. Often, putting up a complete list of possible bugs will help you identify some of them.

Confidentiality, integrity, and availability of information are primary security goals. In a number of contexts, including climate change impact assessment, the term vulnerability has been used.

## *11.2.6 Risk Estimation and Calculation*

Risk evaluation and management focus on the detection of properties, the analysis of vulnerabilities, and the estimation and calculation of potential harm. In general, the risk assessment can be loosely split into qualitative and quantitative evaluations. Qualitative evaluation depends heavily on professional knowledge, while quantitative assessment will measure the system's exact risk benefit. In CPS, researchers have come up with a couple of methods to assess and manage the risks of safety and security; we will be looking into a couple of these methods:

### 11.2.6.1 Safety Risk Assessment Methods

- **Fault Tree Analysis (FTA)**

This method is one of the first, early technology that has been proposed in the safety risk assessment of CPS. Its initial goal is to show the basic possible faulty events that may cause top-level unwanted incidents. This tree has three components: nodes, gates, and edges. The nodes are what we consider the unwanted incidents in the system, gates are the interactions between the nodes, and the edges are the routes of the undesired incidents through our system. Our gates are usually demonstrated with the logical gates, AND gates, and OR gates (Lyu et al. 2019).

- **Failure Modes and Effects Analysis (FMEA)**

FMEA is a coordinated and group-based method used on the system safety analysis. It is used to locate, assess, and mark failures of a risk and the effects. From the name of this method, we can understand that this analysis system looks at the way something could collapse, and the severity of the failure modes (Lyu et al. 2019). Failure modes and effects analysis come in steps, demonstrated in Fig. 11.3.

- **Attack Tree Analysis (ATA)**

This method is somewhat similar to the FTA mentioned previously, but with a slight security difference. This technique takes into consideration the countermeasures of an attack, making an attack countermeasure tree (ACT).

Kure et al. (2018) stated that the Asset Vulnerability Impact Assessment Model (A-VIAM) is considered to analyze the vulnerability influence on the asset. The model

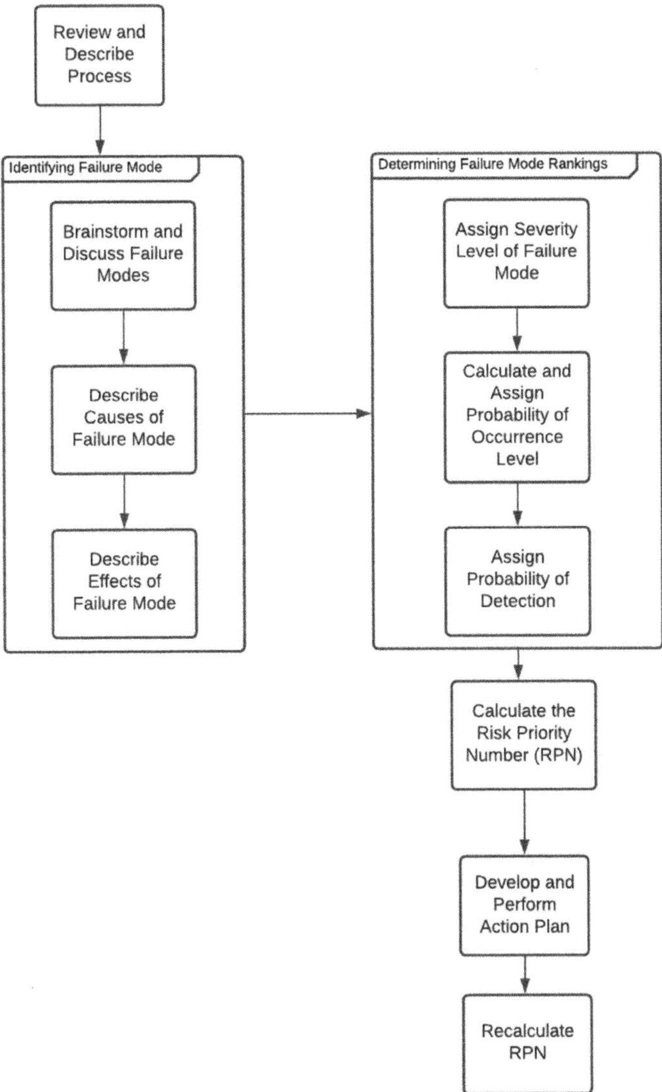


**Fig. 11.3** Steps of Failure Modes and Effects Analysis (FMEA)

is based on the mathematical multi-value theory and is organized as a valuation model. A-VIAM is an integrated assessment technique that assigns a rating on a scale of 0.01–10.0 to the vulnerability effect. Vulnerability Rating (VR) on a scale of 1–5 is used to determine the vulnerability of the essential asset variable and would be separated by the overall amount of vulnerabilities found. The maximum impact value of all critical asset elements will be aggregated and divided by the total amount of critical assets considered to determine the risk of the whole system.

Various vulnerabilities found for the software asset, such as the VR ranking, would be assessed depending on its effect on the software essential asset. The VR values will then be added together to produce an effect value for the program asset and divided by the total number of vulnerabilities found. Such that *VI* = Impacts of vulnerability. Scores range from 1.00 to 10.0 and are assigned to a risk effect on the vital asset. When 1.00–3.99 = low, 4.00–6.99 = medium and 7.00–10.0 = high. *VR* = Ranking of insecurity (Kure et al. 2018). The calculation of the model A-VIAM can be calculated as (Kure et al. 2018)

$$VI(CA) = \sum_{VR=1}^{n} \frac{V_{VR1} + V_{VR2} + V_{VR3} + \cdots + n_{VRn}}{\text{number of vulnerabilities in system}} \tag{11.1}$$

## 11.3 Designing Secure CPS

### *11.3.1 Architecture of the CPS System Overview*

The subsystems include the sensors for sensing the outside environment, the control center where all the computations and logic is performed and information is displayed (Arduino IDE and the laptop), the base station where all the information from the sensors is gathered. Arduino board, the encryption algorithm, and library are used for the security of the information; the wires and USB cable are used to connect the sensor with the control center.

For the subsystems that express security features, we used an encryption algorithm, specifically RC4, to encrypt and decrypt the information gathered from the sensors. It is a variable length key algorithm and a stream cipher that encrypts one bite at a time and it can encrypt one unit at a time. We found an implementation of this algorithm that can be used in the Arduino IDE. We chose RC4 for its simplicity in implementing and using it and the availability of the library in the Arduino IDE. Major advantages of the RC4 algorithm are that it is faster than other ciphers and that it does not need a lot of memory. Additionally, the control system that we used which is a Windows 10 Desktop was installed with antivirus software as well as firewalls. This made the control center more secure against many security threats.

Some subsystems appeared to be more susceptible to vulnerabilities because the sensors appear to be vulnerable to outside attacks such as jamming and denial of service attacks because the algorithm only protects the confidentiality and integrity of the information gathered. However, since it is a wired connection rather than wireless (connecting to the laptop), the attacks can only be performed locally or from the control system itself.

Furthermore, for building security into cyber/physical systems at architectural level we can add a hashing function to protect the integrity of the information gathered. We can also install firewalls on the control system to prevent any denial of service attacks targeting the system.

In addition, If the sensor subsystem is compromised, it could affect the integrity of the system since the attacker can change or affect the information that is gathered by the sensors. This could cause the control system to compute wrong information and send that information to the monitor. The operator could make decisions that can negatively impact the farming and agriculture the system is installed in. This in turn affects the reliability of the system.

Additionally, if the attacker was able to find the key of RC4 and break the cryptography, then this would compromise the confidentiality of the system.

### *11.3.2 Risk of Interaction with External Systems*

Since the system is connected to a laptop which acts as a control system, any malware or vulnerabilities found in the laptop can transfer and affect the system that is directly connected to the smart farming.

### *11.3.3 Architecture Changes Over Time*

For future research, we would try to add more sensors to the board and introduce actuators to the smart farming system. Additionally, for further security, we would add hashing functions and algorithms to the code to protect the integrity of the system. This in turn would affect the code and computation of the code wherein we would add the hashing function and add functions for the additional sensors and actuators. As a result, more security concerns would arise and need to be dealt with.

Table 11.3 will be the table detailing the assets that were determined in the risk assessment and the assets that were actually used.

### *11.3.4 CPS Security Model*

See Figs. 11.4, 11.5, 11.6, 11.7, 11.8 and 11.9.

**Table 11.3** Asset mapping

| Assets | Included or not |
|---|---|
| Base Shield | Yes |
| Arduino Uno | Yes |
| SparkFun soil moisture sensor | Yes, but we used the grove moisture sensor |
| DHT22 temperature sensor | Yes, but we used TMP36 temperature sensor |
| PIR sensor | No |
| Laptop | Yes, but we used a Windows Desktop |
| Data | Yes |
| Code | Yes |
| Network (out of scope) | Yes, using a USB, we connected Arduino with the desktop |

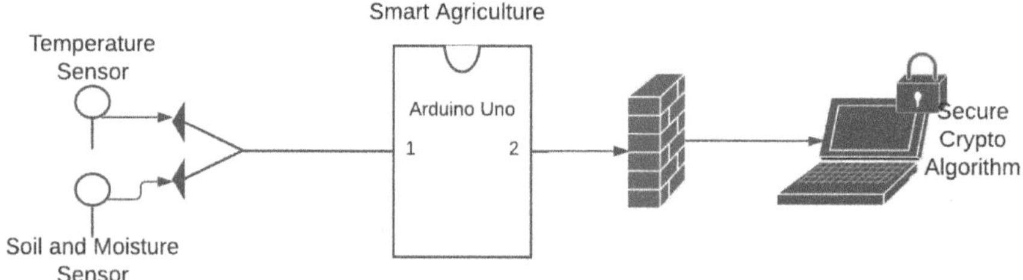


**Fig. 11.4** Smart agriculture flow chart

## *11.3.5 Assets to Security Model Tracing*

There are locations in the architecture where there are no security measures in place that protect the assets. For example, the Arduino board does not have security measures to protect from jamming attacks or denial of service attacks. Also, they are far from view from the operators making them vulnerable to attacks that can go unnoticed.

Furthermore, every location in the architecture has an asset residing there since the sensors are found on the Arduino board and the code and interface can be found in the laptop. Additionally, the cables can be there to connect the CPS to the laptop (control center) for communication between them.

Since the Arduino board acts as the base station, many of the assets including the sensors and the data are found at that location which makes it a target for attackers to try to exploit the vulnerabilities found in the board.

Fig. 11.5 Arduino setup

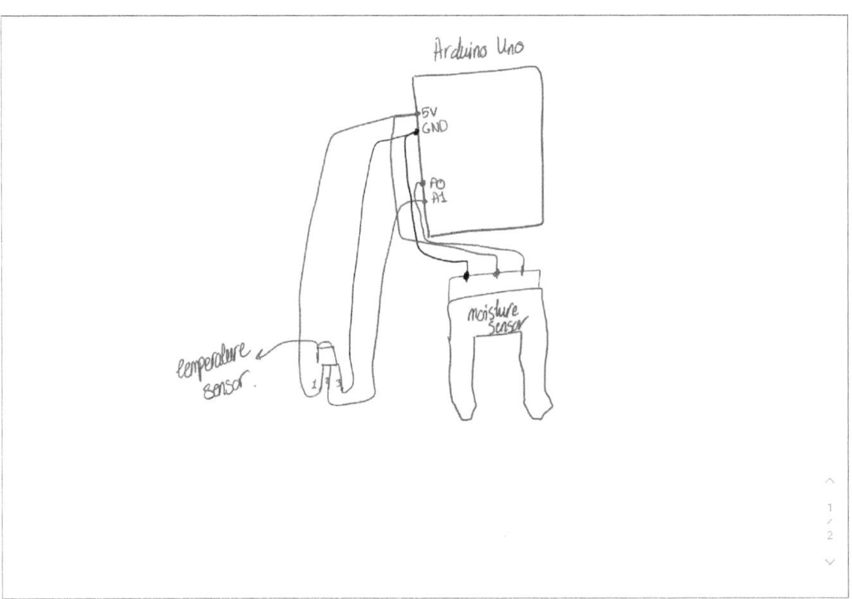


Fig. 11.6 Arduino setup diagram

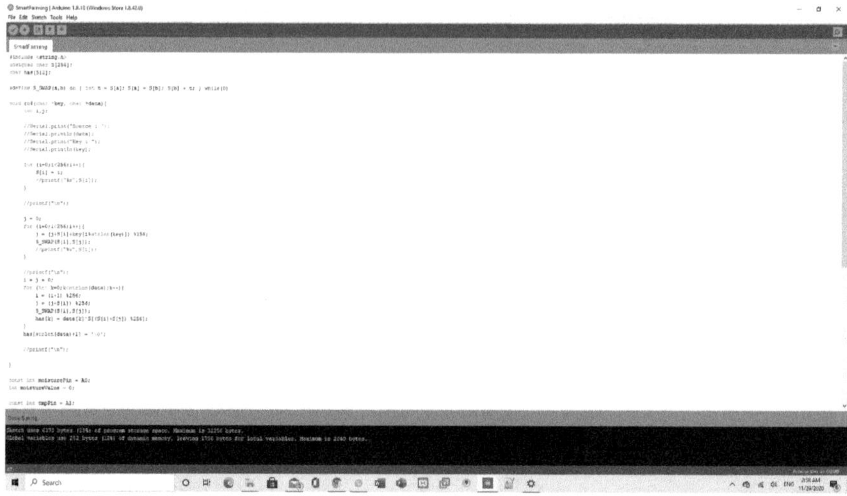

Fig. 11.7 Arduino code 1

Fig. 11.8 Arduino code 2

# 11.4 CPS Security Analysis

## 11.4.1 Analysis of Selection

The three subdomains in aggregation we chose are the code to perform the calculation, the encryption algorithm, and the communication between the aggregator and the

**Fig. 11.9** Code results

base station. We chose those subdomains because most of the security is done in that area and the computation and most of the logic is done there. It is also the area where most attackers would target because it is where the data collected from the sensors is located.

This selection, in our opinion, is good because it allows us to make a full study on the various attacks that can occur to the system since the code and logic are performed at this subdomain. It also allows us to see how effective the encryption algorithm is in protecting the sensitivity of the data collected. Additionally, it gives us an understanding of what the weaknesses are in the algorithm.

## *11.4.2 Security Analysis Results*

These subdomains are where the data collected is stored and where computation on the data is performed. It is also where the encryption on the data is performed. It is an important subdomain where an attacker can perform various attacks which can disrupt the function of the system. These attacks include tampering with the code so that it writes data to the sensors rather than reading from the sensors or even changing which pin to read the data from. An attacker can even try to break the encryption algorithm and gain sensitive information. Additionally, at this area if the laptop or the control center are not securely protected with a firewall, attackers can perform a denial of service attack on the machine and prevent the data from being stored.

The assets at this subdomain that can be affected by security concerns include the laptop connected to the smart agriculture system, the code that performs all the computation and logic on the data collected, the communication between the laptop and system, and the encryption algorithm. These assets are vulnerable to various attacks as mentioned before.

### 11.4.3 Security Risks

If a stakeholder performs a mistake, it could compromise the security of the system itself. It could cause attackers to hijack the system and perform various attacks at this level. This includes tampering with the sensor system and even hijacking the CPS (Cyber-Physical System) and stealing its data. These attacks could lead to critical damage that can cause crops to die and prevent the healthy growth of plants.

### 11.4.4 Potential Vulnerability in the Subdomain

- Information Disclosure where the attacker can break the encryption algorithm and steal the information that is stored in the control center (Laptop). This will compromise the confidentiality of the system.
- Denial of service can occur in this domain, especially if the control center or laptop does not have a firewall installed as a precaution. This is a serious vulnerability because it could prevent the data from being collected and stored by the control system. This will prevent the operator and other stakeholders from analyzing data especially at a critical time in the agricultural domain such as during a rainstorm.
- Man-in-the-Middle Attack can occur as the attacker can intercept the communication between the Arduino board and the control center (laptop) and the attacker can spoof the control system to steal the information being transmitted.
- This subdomain is also vulnerable to a hijacking attack where the attacker can hijack the control system and tamper with the code and logic contained in the control system which compromises the integrity of the system.

### 11.4.5 Project-Specific Checklist

- Check for any denial of service vulnerability in the sensors.
- Protect sensors from any jamming attacks.
- Include hashing function to the code algorithm to protect the integrity of the system.
- Prevent any malicious nodes from entering the cyber-physical system.
- Make sure to replace the moisture sensor as it oxidizes over time.
- Include actuators in the future design.
- Protect actuators from any tampering.
- Provide a backup for all the data collected by the control center.

## 11.5 Conclusion

With the advent of smart IoT devices, their applications have paved their ways immensely into various fields of day-to-day utilities, including agricultural and farming industries. This chapter is a study about how to access the security of the smart agricultural systems. From our findings, analysis, and research we were able to gain a deeper understanding of the many aspects of smart agriculture through our emulation on the Arduino board. From our findings, we found that we were able to protect the sensitive information collected from the sensors using the encryption algorithm, RC4. By applying the encryption, we are able to achieve data confidentiality during data aggregation and dissemination. Although the model used in this chapter is small but can help the young researchers to find a lead into simulation and testing environment. However, since we did not include some integrity checking methods like hash function, the data we collected is susceptible to being tampered. Additionally, we conducted a thorough risk assessment on smart agriculture systems. We conclude that by adapting proactive risk assessments we can reduce the impact of security incidents on the critical assets of smart agriculture system and hence preserve the security requirements. In our future work we propose to extend our model that focuses on end-to-end security and privacy. We also envision to apply encryption on go, during data collection, aggregation, and storage on servers or cloud-based systems.

## References

Gupta M, Abdelsalam M, Khorsandroo S, Mittal S (2020) Security and privacy in smart farming: challenges and opportunities. IEEE Access 8

Kure H, Islam S, Razzaque M (2018) An integrated cyber security risk management approach for a cyber-physical system. Appl Sci 8(6):898

Lyu X, Ding Y, Yang S (2019) Safety and security risk assessment in cyber-physical system. IET Cyber Phys Syst: Theory Appl 4(3):221–232. https://doi.org/10.1049/IET-CPS.2018.5068

# Chapter 12
# Urea Spreaders for Improving the Crop Productivity in Agriculture: Recent Developments

**Deepika Koundal and Virendar Kadyan**

**Abstract** Fertilizers are commonly used for increasing the productivity of crops that, in particular, depends on the fertility of the soil and are estimated by soil test and particular crop. Urea is a solid granular form of fertilizer. Nowadays, it has been observed that with the aid of conventional agricultural practices, farmers are not competent to yield more crops. Farmers have to carry heavy bags, and the spreading process has been performed by hand for small-scale farming in the traditional dispensing of granular fertilizers. The major issues with this method faced by the farmers are urea wastage, uneven spreading of fertilizers, more time-consuming process, and high human effort. This resulted in economic loss and less productivity. Therefore, number of equipment have been developed in literature to solve these issues for small-scale farming. Farmers are able to yield more crop productions with the aid of automatic equipment's for agriculture farming that have an impact on Indian economy. In this paper, some of these automatic urea spreader equipment have been discussed with their pros and cons. By using these spreaders, a lot of time can be saved, human efforts used for carrying heavy bags of fertilizer are reduced, and wastage of fertilizer can also be avoided.

**Keywords** Fertilizer · Fertilizer spreader · Urea · Flow control mechanism · Agriculture farming · Labour cost · Solid Fertilizer · Fertilization time · Spreading

---

D. Koundal (✉)
Department of Systemics, School of Computer Science, University of Petroleum and Energy Studies, Dehradun, India
e-mail: dkoundal@ddn.upes.ac.in

V. Kadyan
Department of Informatics, School of Computer Science, University of Petroleum and Energy Studies, Dehradun, India
e-mail: vkadyan@dd.upes.ac.in



A. Choudhury et al. (eds.), *Smart Agriculture Automation using Advanced Technologies*, Transactions on Computer Systems and Networks,
https://doi.org/10.1007/978-981-16-6124-2_13

## 12.1 Introduction

The Indian nation is mostly based on agriculture. Near about 70% of individuals of our nation are farmers; the Indian economy mainly relies upon agricultural items. The cultivating process incorporates numerous stages. Amongst those cultivation process, fertilization is one of the imperative phases which has not been investigated sufficiently till now. Fertilizers are generally utilized for expanding the productivity of harvests relying upon the soil richness generally estimated by soil tests as indicated by the specific yield (Arjun 2013). Urea is a solid granular type of fertilizer illustrated in Fig. 12.1, which is used to upgrade the development of plants with various application rates relying upon the soil fertility, typically estimated by a soil test and as indicated by the specific yield (Yahya 2018). About 90% of compost is connected as solids (e.g., Urea, Di-Ammonium Phosphate and Super Phosphate). Strong manures are one of the critical hotspots for plant nourishment because of their low cost when contrasted with fluid compound composts. They give imperative supplements to the plants for their development amid its developing life period to enhance the properties of soil (causticity degree and the dirt structure). Spreading the urea unevenly influences the general harvest productivity and the overall profits because of loss of yield quality and yield (Vignesh.B, Navaneetha Krishnan.M, Sethuraman.N 2017). Improper rate of application or non-uniform spreading brought inappropriate urea spreading which necessitates that the ideal rate is resolved and conveyed effectively (Richards and Hobson 2013). Chaudhari et al. introduced the equipment that has a minimal capital cost in comparison to conventional as well as tractor-based fertilizer spreader machines (Chaudhari et al. 2017).

There is a need of automatic option in contrast to the conventional strategies such as tractor based fertilizer spreaders that can satisfy all requirements. Some of the

**Fig. 12.1** Urea Fertilizer (https://www.indiatoday.in/magazine/nation/story/20150202-modi-government-farmers-urea-fertiliser-shortage-817322-2015-01-22)

issues like uneven spreading of the fertilizers (wrong amount and stuff) that may result in damaging of crops are generally faced in conventional fertilizer spreaders. So in order to solve these issues, various spreader machines have been developed for small-scale farming. This leads to an efficient, effective and uniform distribution of fertilizers for the improvement in the production of crops so that the farmers can work efficiently, more easily, and in a functional manner. Farmers are able to yield more productivity with the use of automatic equipment for agriculture farming which ultimately has an impact on the Indian economy (Admade and Jackson 2014).

The contributions of this paper are (i) Number of equipment for the spreading of urea has been discussed. (ii) Pros and cons of different fertilizer spreading machines have been discussed. (iii) Various issues, as well as challenges, have been discussed with the future scope.

Remaining paper is organized as Section 2 has discussed various types of urea spreaders, whereas Section 3 has presented challenges and discussion. Conclusion has been given in Sect. 4.

## 12.2 Different Types of Urea Spreader Mechanisms

An analysis of the literature on the subject shows that to date, only a few issues associated with the process of spreading granular fertilizers with disc spreaders have been explained (Chaudhari et al. 2017; Narode et al. 2015; Fertilizer spreading machine, 2017). There are many types of distribution machines of solid chemical fertilizers. The fertilizer spreading mechanism should satisfy the objectives such as less fertilization process, portability, simplicity of mechanical design, automatically driven, lightweight, wide operation width, low cost, high performance, user-friendly, good distribution accuracy at the desired rate, low power consumption, eco-friendly, ease of maintenance, less time consuming and maximum land cover (Fertilizer spredaing machine, 2017; Boman et al. 2009; Shahaji et al. 2020). Any type of equipment does still not achieve all these objectives. Some of the features are only given in few equipment. Different types of spreaders commonly utilized for the spreading of fertilizers in the farm are handheld, liquid fertilizer, drop spreaders, and rotary fertilizer spreaders. Various types of fertilizer spreading mechanisms have been discussed as follows.

### *12.2.1 Traditional Method*

In conventional methods, fertilizers are spread by hand for small-scale farming, as shown in Fig. 12.2. The major issues with this method faced by the farmers are uneven distribution as well as wrong stuff, and wrong amount of fertilizers may result in crop damage. Moreover, conventional methods of fertilizer distribution in a farm entails high individual efforts and is a more time-consuming process. This takes a lot of time

**Fig. 12.2** Traditional method of urea sprinkling by hand (https://www.indiatoday.in/magazine/nation/story/20150202-modi-government-farmers-urea-fertiliser-shortage-817322-2015-01-22)

to spread the nutrients. The farmers need to hold weighty urea bags along with them during the distribution process. Also, the nutrients are not spread evenly throughout the farm leading to either overdosing or undernourishment to plants in the farm. For small-scale uses, it is not possible to use costly tractor-mounted spreaders (Laghari et al. 2014).

### *12.2.2 Tank Based Fertilizer Spreading Machine*

In this, a machine is developed for spreading granular fertilizers (Liedekerke 2007) over an uncultivated land by dispensing the fertilizer on the spreader circle. The design of this machine consists of three levels: top, middle and bottom level. Hopper is the main component, which is mounted at the top level. Gear arrangement, chain drive and spreader disc are situated at the middle level, and the wheel is mounted at the bottom level. By using this fertilizer spreading machine, an equal amount of fertilizer is spread over. It ensures simple and easy operation. It is more suitable for small and medium farmers, maintenance cost is very less, human effort is saved, and less tiresome as compared to conventional methods and electricity is not required. The demerit is the weight of the tank, which is a concern (Fertilizer spreading machine, 2017).

### 12.2.3 Manually Operated Based Fertilizer Spreader

Another manual operated machine is developed to spread fertilizers uniformly over an uncultivated land using the impeller disc by dropping the fertilizers. In this system, there are three wheels, out of which two are at the front end and one wheel at the back end. The two wheels located at the front end are employed for dispersing the granular fertilizers uniformly over an uncultivated land. The fertilizer is stored with the help of two hoppers, and both the hoppers are situated at some specific height from the wheel's axle in order to make the fertilizer dispense onto the impeller. A flow control mechanism is provided to the hopper, and it is necessary to maintain the flow in fertilization. Spring Mechanism is used to provide a sufficient amount of fertilizer to crops. In typical situations, the hopper is closed and the spring is not in a tension mode. On applying tension by an operator on the spring, the hopper gets open by moving the controlling plate backward. There is one impeller below the system that is mounted upon the output shaft. Impeller is opened by hopper eccentrically and fertilizer get spread in the farm due to the centrifugal action. With the assistance of appropriate gear reduction ratio, this high ratio of centrifugal force is produced. The shaft of wheel is coupled by the gears. With this machine, it is observed that the time as well as the labour cost required for fertilization has been decreased to 50% when compared with the regular approach, which was 80%. It lead to time saving, great speed fertilization, appropriate for small farms, less tiredness to labour, comfortable to operate, easy to assemble, no expert operator is necessary, pollution free, no electric power is required and less maintenance cost.

The limitations of this machine are that it entails further labour in hard land, operational force differs from person to person and manually operated; hence, it is hard to run continuously (Narode et al. 2015). Another inexpensive and simple fertilizer spreader has been introduced. It is the 'walk-behind' appliance that can be quickly and easily driven by farmers and operated manually to spread granular fertilizers such as urea in farms (Deshpande et al. 2018). Motion is transmitted from rear axle wheels on pushing the vehicle which is then transmitted by the rotational motion through chain mechanism and sprocket. Then second sprocket is rotated to connect with a shaft entertaining screw conveyor on both ends. This screw conveyor on rotation dispensed urea through a hopper that is acting as a storage tank (Liedekerke, et al. 2006).

In this, a portable solid-fertilizing apparatus is fabricated for dispensing fertilizer pallets. In this, design, fabricating mechanism and testing of the dispensing mechanism of solid fertilizer have been done. The developed dispenser has been consisted of a cup, which contains fertilizer pallets. The cup is attached below to the storage tank, where a trigger is fabricated for performing the dispensing action of the fertilizer pallets. Hence, fertilizer pallets can be dispensed into a metal pipe through a flexible hose. Cone like sharp pointy edge at the end of the metal pipe that aided the agricultural workers to pierce the soil with ease. It can minimize the time required to perform fertilizing of plants in the agricultural ecosystem and industry. Therefore, there is a huge market for this machine being able to fertilize plants with limited time and force (Guan et al. 2019; Lin et al. 2017).

### 12.2.4 Wheel Based Automatic Fertilizer Spreader

In another one, fertilizer is spreaded automatically over the farms by releasing the granular fertilizers across the impeller disc. A 25 cc engine is used to rotate impeller disc, in which the fertilizer drained and spreaded from hopper, where it is introduced. In tractor mounted or manual system, those carry four and three wheels respectively. However, in this case, two wheels are used in which the bigger front wheel is connected to engine through supporting wheel that can be adjustable. The speed of wheel is varied by control lever connected through a cable. In this, the fertilizer spreaded only in front side of impeller while its back side 180 is covered. The size and width of the fertilizer is reduced to make it less heavy and suitable for multi crops. From this machine, the cost of fertilizer spreader is reduced by 50% (Birajdar et al. 2018).

### 12.2.5 Trolley Mounted Fertilizer Spreader

In this mechanism, the trolley mounted fertilizer spreader is developed. A gear is coupled to the shaft of the wheels of the trolley, which is meshed to another gear in a vertical shaft. The vertical shaft is consisted of a spreader disk, which spreaded the fertilizers to different directions. The handles of the trolley are used to move it forward or backward. As the wheel rotates, through gear transmission the spreader disk rotates. The fertilizer is stored in a vessel at the top. The vessel has an adjustable opening at the bottom that controlled the amount of fertilizer. As the opening is adjusted to the required level, the granular fertilizer is poured into the spreader disk. The fins or vanes in the spreader disk direct the fertilizer to different directions. The fertilizer is spreaded to different directions evenly and in a required amount.

Aravind et al. developed a fertilizer spreader which is based on a trolley type mechanism. The main part used in this mechanism is spreader disk, which helps in uniform spreading. The feed for the disk is from the wheels of the trolley using gear transmission. By using this spreader, a lot of time can be saved, human effort used for carrying heavy bags of fertilizer is reduced and wastage of fertilizer can also be avoided (Aravind et al. 2017). After using it, a uniformity in fertilizer spreading is obtained, human effort is reduced, avoid carrying heavy bags, less wastage of fertilizer, no electrical power is required, economic as well as compact design and reasonable for small farms (Patil et al. 2018). Moreover, larger areas of agricultural land can be covered in a less time and maintenance cost is also very minimal. Hence, it is an economic and compact device, which can be used for fertilizer spreading, especially for small scale farmers. The limitation is that battery maintenance may be required (Aravind et al. 2017).

Laghari et al. discussed the benefits of solid fertilizer in agricultural land. Soil fertilizers consist of several macro and micro constituents that are crucial for crop yield and growth. It is required to keep vital nutrient constituents such as phosphorus, potassium and nitrogen by applying chemical fertilizers (Laghari et al. 2014).

### 12.2.6 Solar Based Fertilizer Spreader

It is studied that in India directly or indirectly 73% of population depends upon farming. The foremost purpose of fertilizer spreader at planting time is to distribute the fertilizers uniformly over the complete farm. The present trend in fertilizer broadcaster in India is based on manual method. It is time to replace the manual method by the motorised. It will decrease the manual effort and time to spread the fertilizer over the entire field. The presented work is concentrated on designing and fabrication of fertilizer spreader, which will use the solar energy to run the motor. This makes the spreading work easier, more efficient and takes less time to spread the fertilizer on farms (Patil et al. 2018; Ramachandhra and Devakumar 2017). Kweon & Grift (Kweon and Grift 2006) have developed a strategy that can control the dropping locale of granular fertilizers onto a spinner disc for optimizing the uniformity in spread pattern. It is based on optical sensor to act as a feedback process that quantified the location and discharge velocity, and diameter of particles for predicting a spreading pattern of a single disc.

Das et al. (Das et al. 2015) reviewed various types of pesticides and fertilizer spreaders like Aerial Sprayer, Lite-Trac, Backpack sprayer, Motorcycle Driven Multi-Purpose Farming apparatus and their pros and cons. Joshua et al. (Joshua et al. 2010) have introduced a solar operated sprayer. With the increased use of engine-propelled pumps, area of irrigated land is increased in the world. The increased cost of oil-based fuel has cut down the profit to be earned by farmers, as cost of food have normally been prohibited from increasing in streak. Adamade et al. (Admade and Jackson 2014) have presented the mechanization that can recognized the necessary major means, which are required to hasten the agricultural production. Sufficiency of food can be achieved by promoting and encouraging the native designs and implementations of equipment at less cost.

### 12.2.7 Tractor Mounted Fertilizer Spreader

Kishore et al. have discussed several mechanisms for farming such as automated preparation of land in which animal-driven or power-driven tools or tractors are used (Kishore et al. 2017). These spreaders are generally of two types: Single disc and twin disc.

a. **Single disc spreaders** are least expensive and easy to develop. It comprised of various spreading blades with single spinning disc. This disc spins in a single direction and hence fertilizer is thrown out on a field. Fertilizers are dropped to provide maximum coverage to crops on both right and left of the machine, and the tractor will also get the same amount of fertilizer quantity, which has to be thrown. Without appropriate diverters to coordinate the fertilizers towards the ground, powder fertilizers create a high level of drift before they arrive on the ground (Przywara 2015).

b. **Twin disc spreaders** are more complex in the design. The development of the unit more often require three gearboxes. Two gearboxes are driven on the left and right side, whereas third gearbox is used as central one, each with spinning disc coordinated to turn it in reverse directions. The constitution of twin discs essentially work in the same way as the single disc spinner does. It can camouflage a broader band, yet fertilizer is still dispersed to provide a uniform spreading. However, the amount of fertilizer released to the left hand side and right hand side can differ by controlling separate pedals. Some of the manufacturers have structured chutes at the back of the spinning discs so that the fertilizers drop on the sides instead of dropping onto the tractor path and these were considerably more successful. Fundamentally, this kind of spinners accomplishes the throwing distance through rapid spinning discs. Cunha et al. discussed that the quality and nature of fertilizer delivery practice is essential for successful spreading in agricultural land. This aimed to study the uniformity in fertilizers spreading with spreaders equipped to perform variable rate. The spreader dissemination of fertilizer by centrifugal spreaders carried out haphazardly over the farms (Cunha and Filho 2016). Thus, the assessment of application rate of fertilizers with right amount should be carried out again for each and every type of product, even on apparatuses with limited variable rate. The advantages are increased uniformity of fertilizer spreading, good crop growth, less time required, less human effort, less waste, eco-friendly. The disadvantages are the volume of hopper that is abridged the repeated refilling of the hopper is mandatory, more frication and lubrication is required (Przywara 2015).

Kshirsagar et al. introduced a multifunctional vehicle for agricultural purpose that can carry several processes like grass eruption from roots, seed bowing and fertilizer spraying (Kshirsagar Prashant et al. 2016). The major issue generally lies with small-scale farms as mechanization is in contradiction of the "economics of scale". These issues are categorized as economic and financial problems, technological constraints and environmental issues. It focused on the problems of seed sowing, fertilizers spraying and grass eruption that are usually faced by fellow farmers. Mada et al. mentioned the importance of mechanization in agricultural farms (Mada and Mahai 2013). The paper focused on the need of simple and cheap vehicle for easiness of several processes in farm. Various fertilizer spreaders are discussed in Table 12.1 with their advantages and disadvantages.

**Table 12.1** Various fertilizer spreaders with their advantages and disadvantages

| Method | Advantage | Disadvantage |
|---|---|---|
| Traditional | Suitable for small scale farming | Uneven spreading<br>Labour some<br>Manually<br>Tiresome |
| Tank based | Equal amount of fertilizer is spread over | Weight of the tank |
| Manually operated | Time saving,<br>Less fatigue to labour,<br>High speed fertilization,<br>No electric power is required | Need more efforts in hard land, manually operated,<br>Operational force differs from person to person |
| Wheel based | Cost fertilizer spreader is reduced by 50% | Not suitable for small scale |
| Trolley mounted | Human effort is reduced, Avoid carrying heavy bags, less wastage of fertilizer, no electrical power is required | Battery maintenance may be required |
| Solar based | Uses the solar energy to run the motor | Maintenance of solar system |
| Tractor mounted | Good crop growth, less time required, less human effort, less waste, eco friendly | The volume of hopper is abridged the repeated refilling of the hopper is mandatory, more frication and lubrication is required |

## 12.3 Challenges and Issues

There is a crucial requisite to automate the operations of agricultural practices (employing machines for carrying out the agricultural practices such as irrigation, fertilizing, weeding, etc.) as it would not only increasing the production as well as decreasing the wastage of fertilizers and saving the labour force. The presented equipment are expensive and not beneficial for small farms.

However, the available machines have various shortcomings such as the machinery is to be operated manually, wastage, over-dosage or under-dosage of the fertilizer (Magó 2020). Moreover, the machines are not capable of detecting the presence or absence of green crop on which the fertilizer/pesticide is to be sprayed. Additionally, the machines used are not portable, eco-friendly thereby causing human fatigue to the farmers during manual operation. Thus, there is a need in the market to provide an automatic fertilizer spreading machine which can handle all these issues. Therefore, an automatic fertilizer spreader is required which should be portable, light weight and have computer vision features. By using automatic fertilizer spreader, a lot of time can be saved, human effort used for carrying heavy bags of fertilizer is reduced and wastage of fertilizer can also be avoided.

The need for agricultural modernization in India must be surveyed by interaction with farmers of small fields. Sustainable enhancement in the occupation of modest agriculturists in emerging nation is mostly dependent on the acceptance of enhanced technologies for yielding better crop production. As number of technologies are already existing but their accessibility and execution is lacking. Moreover, the interaction between agricultural research and farmers are also deficient.

Almost all farmers still use simple methods of farming due to lacking of financial support or lacking of knowledge to utilize current equipment. The utilization instruments used by hands for farming is still prevalent in India as tractors entail assets that numerous Indian farmers do not have easy access too. With the adoption of scientific farming techniques, farmers can acquire high yield and great quality of crops, which can keep the farmers from going impoverish.

## 12.4 Conclusion

Agriculture is the main foundation of an Indian economy and will remain to be for a longer time. Modern agricultural methods and equipment have not been utilized by small scale holders because of the apparatus cost and hard to obtain. This paper discussed various issues of farmers experiencing the increase in labour cost for fertilizing. The disadvantage as well as advantages of existing spreader models have been examined and discussed. From this survey, it has been observed that the thriftily feebler farmers will be benefitted by the semi-automatic fertilizer spreaders that can efficiently encounter the nourishing necessities of small farms. This would result in the rise in worthy crop yield, fewer human exhaustion, least usage of fertilizers, and uniformity in urea spreading, decline in time required to spread and less wastage at minimum cost in contrast to current available machines. Further, research and development can be performed on a mechanism that can control the discharge of fertilizers in a better way that can be obtained by flow control valves such as butterfly valve and ball valve. Nowadays, it is manual operated and can be made power driven by fastening small battery or can be made solar operated. Future work can be extended to incorporate sensor based, drone based and computer vision based systems to decrease the human exertions, to extend the continuous work limit, to enhance the productivity and less destructive to user. Moreover, the hopper's capacity can be improved in order to cover large field with least fill-up prerequisite without intensifying the machine weight.

**Acknowledgements** Thanks to UPES for supporting in this research.

**Funding:** Seed grant has been obtained from UPES, Dehradun.

**Consent for Publication:** All authors have their consent in publishing.

**Conflict of Interest:** No conflict of interest.

## References


Admade CA, Jackson BA (2014) Agricultural mechanization: a strategy for food sufficiency. Sch J Agric Sci 4(3):152–156

Aravind MG, Balan AS, Abraham A, Sajeev A (2017) Design and fabrication of trolley mounted fertilizer spreader. Int J Innov Res Sci-Ence Technol 3(11):252–254

Arjun KM (2013) Indian agriculture-status, importance and role in Indian economy. Int J Agric Food Sci Technol 4(4):343–346

Birajdar BR, Awate MB, Kulkarni PN, Dhobale AD (2018) Design and fabrication of fertiliser spreading machine. Int J Res Appl Sci Eng Technol 6(6):238–245

Boman B, Cole D, Futch S, Gunter W, Wilson C, Hebb JW (2009) Fertilizer application. Best management practices for citrus grove workers. http://edis.ifas.ufl.edu

Chaudhari S, Naeem M, Jigar P, Preyash P (2017) Design and development of fertilizer spreader machine. Int J Eng Sci Res Technol 6(4):62

Cunha JP, Filho RS (2016) Broadcast distribution uniformity of fertilizer with centrifugal spreaders used in variable rate application. Engenharia Agrícola 36(5):928–937

Das N, Maske N, Khawas V, Choudhary SK (2015) Agricultural fertilizers spreaders and pesticides-a review. Int J Innov Res Sci Technol 1(11):44–47

Deshpande PM, Shinde MA, Wadile AS, Vaidya RR, Shukla VP (2018) Design and development of manually operated fertilizer spreader. 6(3):2402–2406

Guan CC, Tian KS, Lin SC, Singh SS (2019) The concept of solid fertilizer dispensing machine. Politek Kolej Komuniti J Eng Technol 14:40–51

https://www.indiatoday.in/magazine/nation/story/20150202-modi-government-farmers-urea-fertiliser-shortage-817322-2015-01-22

Joshua R, Vasu V, Vincent P (2010) Solar sprayer - an agriculture implement. Int J Sustain Agric 2(1):16–19

Kishore N, Gayathri D, Venkatesh J, Rajeshwari V, Sageeta B, Chandrika A (2017) Present mechanization status in sugarcane–a review. Int J Agric Sci 9(22):4247–4253

Kshirsagar Prashant R, Ghotane K, Kadam P, Arekar O, Insulkar K (2016) Modelling and analysis of multifunctional agricultural vehicle. Int J Res Advent Technol 4(1):53–57

Kweon G, Grift TE (2006) Development of a uniformity controlled granular fertilizer spreader. American Society of Biological and Agricultural Engineers (ASBAE) 2(4):1–14

Laghari M, Laghari N, Shah AR, Chandio FA (2014) Calibration and performance of tractor mounted rotary fertilizer spreader. Int J Adv Res (IJAR) 2(4):839–846

Lin SC, Meng LS, Guan CC (2017) Design Development of Mobile Fertilizer Dispenser (MDF). Jurnal Kejuruteraan Teknologi Dan Sains Sosial 1(1):56–62

Mada DA, Mahai S (2013) The role of agricultural mechanization in the economic development for small scale farms in Adamawa state. Int J Eng Sci 2(11):91–96

Magó L (2020) Smart control on agricultural machines. Hung Agric Eng 37:41–47

Narode RR, Sonawane AB, Mahale RR, Nisal SS, Chaudhari SS, Bhane AB (2015) Manually Operated Fertilizer Spreader. Int J Emerg Technol Adv Eng 5(2):369–373

Pratik PP, Suyog SP, Sumit M, Sonwane (2018) Solar operated solid fertilizer dispensing machine. In: Proceedings of Emerging trends in engineering and technology (NCETET-2018), held at Kolhapur, pp 468–472

Patil PP, Ghodke AS, Kalme SR, Devshette AR (2019) Urea Spreader Machine. Int Res J Eng Technol (IRJET) 6(3):2345–2348

Przywara A (2015) The impact of structural and operational parameters of the centrifugal disc spreader on the spatial distribution of fertilizer. Agric Agric Sci Procedia 7:215–222

Ramachandhra S, Devakumar MLS (2017) Design and Implementation of solar fertilizer broadcaster. Int J Tech Innov Mod Eng Sci 3(12):244–259

Richards IR, Hobson RD (2013) Method of calculating effects of uneven spreading of fertiliser nitrogen. In: Proceedings-International Fertiliser Society. No. 734. International Fertiliser Society

Swapnil B, Arbaz I, Sail S, Suraj, Dighe MD (2018) Fertilizer spreading machine. Int J Adv Eng Res Dev Technophilia -2018 Feb 5(4):1–3

Swapnil B, Arbaz I, Sail S, Suraj Y, Dighe MD (2018) Fertilizer spredaing machine. Int J Adv Eng Res Dev 5(4):1–3

Shahaji KS, Madhukar MK, Ankush BS, Babasaheb DD, Pawar M (2020) Design and development of fertilizer feeder machine. 7(5):113–119

Van L, Paul et al (2006) A discrete element model for simulation of a spinning disc fertilizer spreader I. Single particle simulations. Powder Technol 170(2):71–85

Van Liedekerke P (2007) Study of the granular fertilizers and the centrifugal spreader using Discrete Element Method (DEM) simulations. Catholic University of Leuven, Belgium, Bio-ingenieurswetenschappen

Vignesh B, Navaneetha Krishnan M, Sethuraman N (2017) Design & fabrication of automatic fertilizer spreader. IJSRD-Int J Sci Res Dev 5(4):1133–1135

Yahya N (2018) Efficacy of green urea for sustainable agriculture. Green Urea. Springer, Singapore, pp 99–123

# Chapter 13
# Agricultural Informatics and practices—The Concerns in Developing and Developed Countries

**P. K. Paul**

**Abstract** Agricultural Informatics is an interdisciplinary field of study and has wider potentiality in developing and modernizing overall agricultural systems. Agricultural Informatics is further needed in developing allied areas viz. food and ecological systems as well, partially. Agricultural Informatics is the need of hour for development of the smart agricultural systems. It is a practicing area and also a field of study. Agricultural Informatics is initially only common in the developed countries but in recent past the scenario is been changed as many developing countries also have updated ICT in Agriculture and allied areas. Agricultural Informatics initially only deals with the documentation and information management using computer in the fields of agricultural sectors, but gradually other technologies have been used and increasing day by day. Furthermore the aspects of Management, Statistics, and Societal environment have been added into the Agricultural Informatics for further practice and development. Agricultural Informatics though has variations depending upon the countries and their status. The developing countries are having different views, status in respect of the Agricultural Informatics practice and awareness, whereas the countries from developed one are attributed different featured. This paper talks about aspects of Agricultural Informatics in respect of developed and developing countries as a special focus.

**Keywords** Agricultural informatics · ICT4D · ICT in agricultural · IT · Computing · Development · Developed countries · Digital divide

## 13.1 Introduction

Agriculture plays an important concern in economic and social development in many countries especially in developing and undeveloped countries. It deals with the issues and concern of the food security, health of people, and also in enhancing food quality improvement. Agricultural is needed in every country for the fulfillment

---

P. K. Paul (✉)
Department of CIS, & Information Scientist (Offg.), Raiganj University, Raiganj, India

 
A. Choudhury et al. (eds.), *Smart Agriculture Automation using Advanced Technologies*, Transactions on Computer Systems and Networks,
https://doi.org/10.1007/978-981-16-6124-2_14

of the demand for food including in the poverty and in reduction of the malnutrition. It is worthy to note that agricultural sector also needed to achieve the aspects on Sustainable Agriculture by attending sustainability and with the concern of the natural environment protection. In many a times farmers faced various aspects viz. profit margins, costs of the fertilizers and fuel, product prices including in increased globalization especially in the developing countries. In this respect the mechanisms for Agro IT including other technological transfer under traditional and changed conditions (Abbasi et al. 2014; Goraya and Kaur 2015; Paul 2013).

Previously significant transformation of agricultural sector has been driven by demand, but current information is considered as most important and valuable. Moreover information is required to the farmers and others to reach end-users very fast and to get wider opportunities and benefits. With the help of Agricultural Informatics it is possible to get about the seed, climate, plant protection, market of the crops, and agricultural products. Therefore in successful planning of the farming, Agricultural Informatics plays a leading role. Information-intensive and technology blended farming are the need of the hour keeping in mind the aspects of sustainable agricultural production. Though it is a fact that farmers should be aware about the aspects of the technological implementation specially uses including basics of internet operation with information service systems for healthy management of agricultural production. Further in the economic potentiality enhancement also ICT uses are considered as important. In developing precision farming such as in livestock management also ICT cold be helpful for the basic farmers, managers of the agricultural industries, policy makers and so on (Ahmad et al. 2015; Holster, et al. 2012; Teye et al. 2012).

Agricultural Informatics is also important in enhancing the productivity of the Agricultural products with the support of real-time information which is decision making, broadband, internet, e-mail having pictures, video clips and sounds, etc. These applications are important in operation management of agricultural production. Agricultural Informatics is used as a tool for access to market information, weather and climate, technology and so on. All these technological systems are responsible in making advantages to the farmers for the effective agricultural production and obviously improvement of their life quality. There are various challenges, issues and concerns in this regard.

## 13.2 Objective

The present paper entitled 'Agricultural Informatics and Practices—The Concern in Developing and Developed Countries' is a comprehensive overview of the following:

- To learn about the basic of the Agricultural Informatics with reference to its basic operation and activities.
- To know about the major development and challenges in respect of the Agricultural Informatics.

- To learn about the Agricultural Informatics context in the developing and undeveloped countries.
- To find out the issues and scenario of Agricultural Informatics in developed countries and territories.
- To know the digital divide in the context of Agricultural Informatics and related concern of the same.

## 13.3 Agricultural Informatics: Inception and Context of Developing and Developed Countries

Agricultural Informatics is significant in agricultural sector both directly and indirectly. ICT is directly helpful in enhancing the productivity of agricultural sector and indirectly, with the help of Agricultural Informatics, farmers can get the information about various aspects of the agriculture and that can be helpful in further decision making with efficient management of agricultural production and similar enterprises.

Precise farming is very much important and needed in developed countries and with the use of ICT to agricultural productivity will increase rapidly (Gill et al. 2017; Paul et al. 2016; Zamora-Izquierdo et al. 2019). Agricultural Informatics is considered valuable in remote based operations and management also using remote sensing and satellite technology. Here GIS technologies help a lot in not only agriculture but also in environmental and soil developed. ICT integrated Agricultural Informatics practice is helpful in knowing weather condition, and connected to computers is helpful to get information on weather, temperature, soil, rainfall, humidity, moisture, moisture, length of day, wind speed and so on. Therefore in precise farming including in corporate and big agro farm development also Agricultural Informatics is considered as important and valuable. Agricultural Informatics is thus appropriate for corporate farming apart from the small enterprises related with the agriculture. With the help of ICT farmers can get the need timely and reliable information including its sources and that leads future development of agriculture. At present in many countries farmers are depend on conventional information systems which mostly not restricted with the timely information and as a result farmers face various problems and issues though the use of Agricultural Informatics they can stay competitive and survive on globalized market.

All would be consider as waste if farmers are unable to use Agricultural Informatics properly. Therefore for knowing the basic operations of the IT including internet services such as searching useful information on agriculture basic internet, computing and digital literacy is required. Through the internet the farmers can get the information about agro products as described previously and also communicate with colleagues around world and by this information gathering and even exchange of ideas on agricultural products, cultivation of the plants etc. including animals become easy. ICT and Agricultural Informatics are therefore having the contribution to national economy and society (Babu et al. 2015; Balamurugan et al. 2016; Kajol and Akshay 2018).

### 13.3.1 ICT in Agriculture in European Union

The uses of IT and Computing (i.e. Agricultural Informatics) in agricultural sector in EU and Serbia is difficult to gather though country reports regarding the agricultural firms, and farmers can be helpful in gathering information on various technological aspects viz. internet access in different countries. According to the report in January 2009 about 93% of all the enterprises in the EU was having the services of internet and Mobile internet connections were 28% in the agricultural organizations in European Union (Adão et al. 2017; Guardo et al. 2018; Rezník et al. 2015). However according to the report the lowest percent mobile based internet user countries are Greece, Cyprus and Romania.

Though in recent past the mobile communication infrastructure developed significantly, here the usage of mobile and smart phones having various web applications are also increased. Smihily, Storm, 2010 in the study shows that above 90% of farmers in few countries viz. Italy, Ireland, Spain use internet access by mobile though by PC is relatively low. Even mobile devices in such countries are used in higher manner for the phone call and message (SMS). They further expressed the usage of mobile application for business purposes of 2–5% in such countries.

Agricultural Informatics is perfectly suitable for the Precision farming, and it will enhance the automation in agriculture. In many European Union countries precision agriculture is initialed at least in small manner by the concerned farmers and also by the government in certain cases. Here Geospatial data are collected by some of the EU countries. According to the Teye et al., 2012; there are major differences between the countries across Europe. And major differences are noted in Western and Northern Region; for example the Czech Republic is considered important in the progress of Precision Farming. Whereas agricultural machines are well designed and improved in the countries like Germany, Denmark, Netherland and Finland.

Researchers have shown that in the Europe the data integration in intra and inter-enterprise level is not so strong; and the availability of internet in rural areas is an important issues; though few countries are doing proper steps in the standardization viz. Germany, France, Denmark, Belgium and Netherland. Though, in some of the countries like Slovakia, Romania, Lithuania and Czech Republic etc. the private and public engagement in advanced ICT infrastructure is low but gradually improving. According to the expert in the countries like Slovakia, Bulgaria, Italy due to the lack of young people in agriculture, there is a lack of Agricultural Informatics practice. Countries with small and poor farmers may have some problems in farm automation; but even then also few are doing well in agricultural ICT practice like Baltic States and it is due to they have not legacy of old systems. The broadband internet in rural areas in some of the European countries is considered as important factor; and this may be considered as major problem in agriculture. The study of level of ICT in agricultural development in the EU countries and Switzerland shows various aspects and here the same is depicted in Table 13.1 (*Source*: Teye et al., 2012) (Hilbert 2016).

**Table 13.1** Level of ICT and agricultural technology use in the EU countries and Switzerland

| Country | Farm PC | Inter- net | Farm Info. Sys | Phone | LPIS rele-vance | Geo-fertilizing | Animal registra-tion | Data exchange level of develop |
|---|---|---|---|---|---|---|---|---|
| BGR | Low | Low | Low | – | Average | – | – | **Hardly any** |
| CZE | High | High | High | Low | Average | Average | – | **Averagely** |
| DNK | High | High | – | High | High | Average | High | **Well** |
| EST | High | High | Ave-rage | – | Average | Low | Average | **Poorly** |
| FIN | High | High | High | High | High | Average | High | **Well** |
| FRA | High | Ave-rage | Ave-rage | High | High | Average | High | **Well** |
| HUN | Ave-rage | Ave-rage | Low | Low | Average | Low | Average | **Poorly** |
| ITA | Ave-rage | Low | Ave-rage | Ave-rage | Average | Average | High | **Average** |
| LVA | Low | High | Low | – | Average | Low | High | **Poorly** |
| NLD | High | High | High | High | High | Average | High | **Well** |
| POL | Ave-rage | Ave-rage | Ave-rage | – | Average | Low | Average | **Hardly any** |
| ROM | Low | Low | Low | Low | Average | – | Average | **Hardly any** |
| SVK | High | Ave-rage | Low | Low | Average | Low | Average | **Poorly** |
| SVN | Low | Low | Low | Low | Average | – | Average | **Poorly** |
| ESP | High | – | Ave-rage | Low | High | Low | High | **Averagely** |
| SWE | High | – | – | High | High | Low | High | – |
| CHE | **High** | **Ave-rage** | **Ave-rage** | **Low** | **Average** | **Low** | **High** | **Averagely** |

Here most of the data provided with the High, Average, Low, Poor etc. Whereas, Fig. 13.1 is depicted about the percentage of the technological uses according to the concerned study (Channe et al. 2015; Muangprathub et al. 2019; Ojha et al. 2015).

The agricultural sector is suffering with the major challenges viz. increasing production of the feed, decreasing of the availability of natural resources, water shortages, declining soil fertility, climate change, the process of urbanization, production of the higher quality products, and communication of the rural communities; and so on. For improvement of such challenges and issues, proper policy planning standards and regulations including Agricultural Informatics are highly required.

New approaches including technical innovations are also needed in enhancement of the ICT in Agricultural space. Food security is an important issue and concern, and it is been endorsed at the World Summit on the Information Society during the 2003–2005. Here for healthy and developed Agricultural Informatics practice initially following may be considered as steps among the community specially engaged in the agricultural professions; directly and indirectly

- Computers with internet,

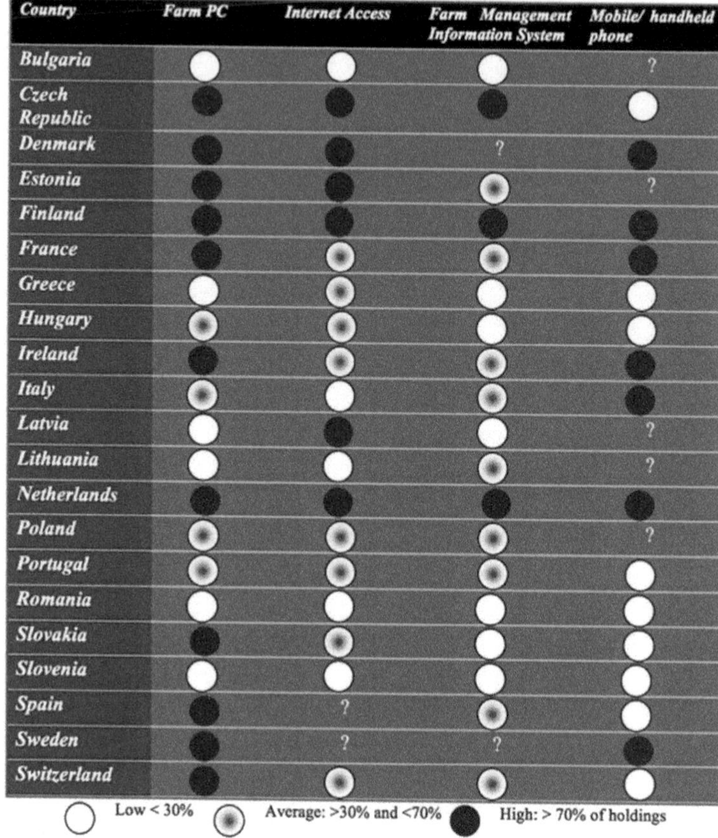

| Country | Farm PC | Internet Access | Farm Management Information System | Mobile/ handheld phone |
|---|---|---|---|---|
| Bulgaria | ○ | ○ | ○ | ? |
| Czech Republic | ● | ● | ● | ○ |
| Denmark | ● | ● | ? | ● |
| Estonia | ● | ● | ◉ | ? |
| Finland | ● | ● | ● | ● |
| France | ● | ◉ | ◉ | ● |
| Greece | ○ | ◉ | ○ | ○ |
| Hungary | ◉ | ◉ | ○ | ○ |
| Ireland | ● | ◉ | ◉ | ● |
| Italy | ◉ | ○ | ◉ | ● |
| Latvia | ○ | ● | ○ | ? |
| Lithuania | ○ | ○ | ◉ | ? |
| Netherlands | ● | ● | ● | ● |
| Poland | ◉ | ◉ | ◉ | ? |
| Portugal | ◉ | ◉ | ◉ | ○ |
| Romania | ○ | ○ | ○ | ○ |
| Slovakia | ● | ◉ | ○ | ○ |
| Slovenia | ○ | ○ | ○ | ○ |
| Spain | ● | ? | ◉ | ○ |
| Sweden | ● | ? | ? | ● |
| Switzerland | ● | ◉ | ◉ | ○ |

○ Low < 30% ◉ Average: >30% and <70% ● High: > 70% of holdings

**Fig. 13.1** Application of ICT (Farm PC, Internet access, Farm management system, Mobile/handheld phone) in agricultural holdings in (most) EU member states

- Basics of geographical information systems,
- Use of mobile phones with use of modern apps may be agriculture oriented.

Apart from these the traditional media such as FM, radio, and TV may also have different role in ICT regarding the uses of agricultural development. The report of the IICD shows various aspects of Agricultural Informatics promotion as well. Further the organization also involved in wide range of projects and policy, impacting ICT utilizations (Adetunji and Joseph 2018; Gómez-Chabla et al. 2019; Paul et al. 2020).

In other sense to reach wider audience, rural radio, TV, FM, and mobile phone could be consider important and valuable. Govi Gnana project in Sri Lanka deals with the project of displaying prices on light boards at major markets of agro commodities and products. The Web-based trading platforms are also emerging, especially for main commodities throughout the world. In India various private and government

**Fig. 13.2** People interested in viewing VSAT in Ghana

initiatives are important to note. Among the important concern of these most valuable are as follows:

- Agriwatch (www.agriwatch.com);
- eChoupal programme (www.itcportal.com/ruraldevp_philosophy/echoupal.htm).

With the support of these, million farmers can get information about the agricultural products, crops, agricultural market by the computing services, email, apps, and also traditional text messages. Moreover in other the so-called less developed and undeveloped countries also ICT uses in Agriculture can be noted for the betterment of the farmers viz. Senegal, Benin, and Zambia. The market information systems project of IICD in Bolivia is also important to note. Similarly the initiative of the same can be noted at the Uganda, Tanzania and Ghana for fostering ICT to Agriculture. IICD supports the Social Enterprise Foundation of West Africa (SEND), Ghana for the linking of rural soybean producers to mills. And interestingly here the uses of satellite technology, database systems, computing with mobile phone uses are important to note (Aubert et al. 2012; Bauckhage and Kersting 2013; Nayyar and Puri 2016). A sample picture is provided in Fig. 13.2.

### *13.3.2 Community Empowerment Initiatives*

Communities and farmer organizations are these days also doing well in the uses of ICT in the agriculture and systems. Such organizations are engaged in community development for strengthening their capacities so that farmers and common people should be aware about the output prices, land claims, resource rights, and infrastructure projects. With the proper initiative of the community empowerment of the local communities, it is helping in national as well as in global developments and allows effective uses of the Agro ICT. Global Positioning Systems inked to Geographical Information Systems are helpful for the rural communities to document and communicate their situation. For example the India Dairy Company Amul India initiated AMUL program that automates milk collection and payments for its 500,000 members; therefore it is much more transparent and ensuring fair payments to farmers and cultivators as well.

However it is important to note that there is a huge gap of knowledge and information among the rural communities due to lack of interest, less infrastructure, awareness, and also less availability of agricultural knowledge centers in the rural communities. According to the expert, it is expected that multi stakeholder mechanisms should be considered as important for the information accessibility to the end users; here rural communities may be benefited with the available knowledge. In India and Chile's online advisory service, mobile Q&A services may also be considered as important in this context. Proper initiative is needed at national level, to ensure learning and information sharing; in this regard it is worthy to note that IICD supports many countries in removing digital divide and here ICT4D networks can be consider as important for the knowledge sharing, togetherness, policy dialogue, and sharing as well (Gómez-Chabla et al. 2019; Kamble et al. 2020; Paul et al. 2015).

The type of ICT used by local communities is subject to rapid change. However, broadband internet access is seen as central for societal innovation because storing of large datasets and live communication requires good connectivity. Until recently, connectivity in rural areas was limited to slow dial-up lines. Satellite connections now make broadband access possible in remote areas. In the rural areas in Africa also these days the uses of the mobile phone increased in recent past. Though there is a difference in broadband access between developed and developing countries, due to the shortage of broadband users, the use of new technologies such as MESH and WiMAX can be a noted problem at African countries. However the initiatives are important to note. The new generation mobile phone networks comes with speed internet services at sharply reduced costs, and therefore in rural areas this can be accepted well even in undeveloped countries as well. The Food and Agricultural Organizations (FAO) of the United Nations and The International Telecommunication Union (ITU) have been engaged for the promotion of ICT in agricultural systems and to lead the E Agriculture internationally. According to the FAO & ITO the significance of the ICT in Agricultural Systems are many and such are mentioned in Fig. 13.3.

The major issue in Digitalization of the Agriculture is lack of infrastructure and availability of basic Computing and Internet Operation. Internet is fruitful for various basic Agricultural Information Technological Operations; with this the farmers can be able to find out the basic information without huge amount of technological support, mechanism, and systems (Adão et al. 2017; Liu et al. 2019; Tsekouropoulos et al. 2013). There is a significant change internationally in terms of internet users in last 10 years. According to the ITU Data sources in the year 2010 the worldwide internet user was 30% of the total population and gradually in recent past the amount

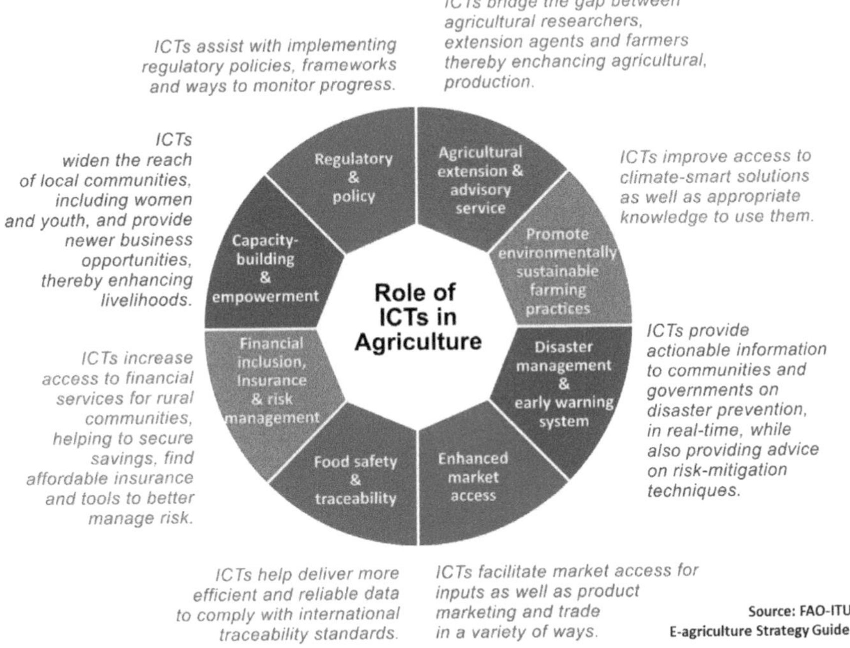

ICTs assist with implementing regulatory policies, frameworks and ways to monitor progress.
ICTs bridge the gap between agricultural researchers, extension agents and farmers thereby enchancing agricultural, production.
ICTs widen the reach of local communities, including women and youth, and provide newer business opportunities, thereby enhancing livelihoods.
ICTs improve access to climate-smart solutions as well as appropriate knowledge to use them.
Regulatory & policy
Agricultural extension & advisory service
Capacity-building & empowerment
Promote environmentally sustainable farming practices
Role of ICTs in Agriculture
Financial inclusion, Insurance & risk management
Disaster management & early warning system
Food safety & traceability
Enhanced market access
ICTs increase access to financial services for rural communities, helping to secure savings, find affordable insurance and tools to better manage risk.
ICTs provide actionable information to communities and governments on disaster prevention, in real-time, while also providing advice on risk-mitigation techniques.
ICTs help deliver more efficient and reliable data to comply with international traceability standards.
ICTs facilitate market access for inputs as well as product marketing and trade in a variety of ways.
Source: FAO-ITU E-agriculture Strategy Guide


**Fig. 13.3** Agro ICT context by FAO-ITU

reach significantly at 53.6%. However there is a significant difference of this data among the developing and developed world. In the year 2010 only 21% of the total population of the developing world was internet connections whereas this data of the same year in developed world was 67% of the total population. Table 13.2 in this regard provided other allied data also including changes of world population since 2005 and user base of the internet in the world, developing world, and developed world (Source ITU report & Wikipedia).

However internationally in the context of continents and regions all the areas and regions are not equally developed and having good stands in respect of internet user. Table 13.3 in respect of this provided a lot (Source ITU report & Wikipedia).

**Table 13.2** Worldwide internet users

| Worldwide Internet users | | | | |
|---|---|---|---|---|
| | 2005 | 2010 | 2017 | 2019[a] |
| World population | 6.5 billion | 6.9 billion | 7.4 billion | 7.75 billion |
| Users worldwide (%) | 16 | 30 | 48 | 53.6 |
| Users in the developing world (%) | 8 | 21 | 41.3 | 47 |
| Users in the developed world (%) | 51 | 67 | 81 | 86.6 |

[a] Estimate. *Source* International Telecommunications Union

**Table 13.3** Worldwide internet users based on regions of the world

| Internet users by region | | | | |
|---|---|---|---|---|
| | 2005 (%) | 2010 (%) | 2017 (%) | 2019[a] (%) |
| Africa | 2 | 10 | 21.8 | 28.2 |
| Americas | 36 | 49 | 65.9 | 77.2 |
| Arab States | 8 | 26 | 43.7 | 51.6 |
| Asia and Pacific | 9 | 23 | 43.9 | 48.4 |
| Commonwealth of Independent States | 10 | 34 | 67.7 | 72.2 |
| Europe | 46 | 67 | 79.6 | 82.5 |

[a] Estimate

Based on above figure it can be easily understandable that the continents and regions like Africa is sharing 28.2% of the internet users of the total population of the respective regions in the latest study of 2019 whereas the statistics of Asia and Pacific is about 48.4% of the total population of the total population of the concerned regions. Though statistics of the two other contents, i.e., North and South America (Americas) is about 77.2% in respect of the total population. The areas of Arab, Commonwealth of Independent States, are having different statistics and depicted in Table 13.3.

The broadband infrastructure is also consider as important and valuable for major Agricultural Information Technology operations (Gill et al. 2017; Khattab et al. 2016; TongKe 2013). The statistics of the broadband is not fine in all the areas and far different in developed and developing world. Further the statistics of the Fixed and Mobile broadband is also different. According to the latest 2019 data the developed countries are having good fixed broadband and that is 33.6% of the total users whereas in developing countries it is only 14.5%. Further in mobile broadband also the statistics are noticeable and details are depicted in Table 13.4 (Source ITU report and Wikipedia).

**Table 13.4** Worldwide broadband subscribers of the world

| Worldwide broadband subscriptions | | | | |
|---|---|---|---|---|
| | 2007 | 2010 | 2016 | 2019[a] |
| World population | 6.6 billion | 6.9 billion | 7.3 billion | 7.75 billion |
| Fixed broadband (%) | 5 | 8 | 11.9 | 14.5 |
| Developing world (%) | 2 | 4 | 8.2 | 11.2 |
| Developed world (%) | 18 | 24 | 30.1 | 33.6 |
| Mobile broadband (%) | 4 | 11 | 49.4 | 83 |
| Developing world (%) | 1 | 4 | 40.9 | 75.2 |
| Developed world (%) | 19 | 43 | 90.3 | 121.7 |

*Source* International Telecommunication Union

## 13.4 Africa, Digital Divide, and Agro Informatics

The digital divide is higher in the countries of Sub-Saharan Africa due to widespread poverty. According to the study only 7% of the continent's inhabitants that are online; though mobile phone users are higher at 72%, however further only 18% of these phones are smartphones and there are lack of internet usage. However, to reduce the digital divide in African region the initiative of the awareness on infrastructure, investment, innovation, entrepreneurs engagement need to address. The increasing rate is higher and noticeable for example in Africa only 2% of the internet users was in the year 2005 and in the year 2010 it was 10%. The significant changes are noticeable in the year 2017 data as it has reached to 28.2% of the total population of Africa.

It is worthy to note that in Africa still a disparage between Internet access of the poor and wealthy peoples due to certain reasons. This becomes an important fact that most of the children are studied in the sub-standard academic institutes except few who are associated with the rich and healthy institutes. Therefore in most cases the most studied candidates are become associated with the technological skill sets only after completion of their academics at their workplace. This is attributed poor infrastructure and technology savvy attitude as well. According to a study it has noted that in 2000 total telephone lines was higher at Manhattan than that of entire Sub Saharan Africa. Such poor infrastructure in Africa also partially reason for the lack in economic development. The total internet access of Africa and whole world is visible with Fig. 13.4 (Source ITU Report) (Babu et al. 2015; Manos et al. 2011; Paul et al. 2015).

According to 2011 estimates, about 13.5% of the African population has Internet access; it is big population zone of the world and accounts for 15% of the World population. And it is also noted that among the countries of Africa, South Africa

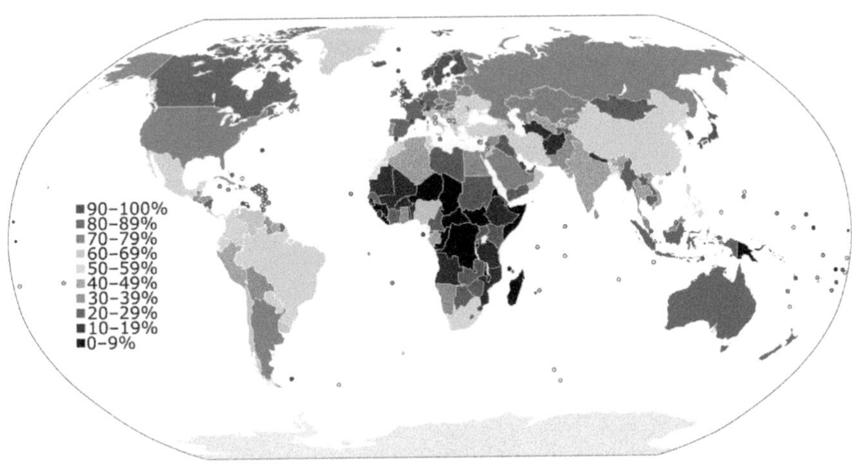


**Fig. 13.4** Internet users percentage based on populations

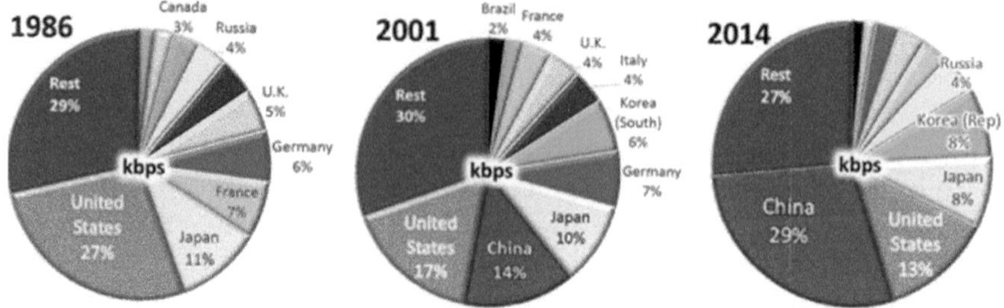


**Fig. 13.5** Provided the top countries in respect of most bandwidth of the past

infrastructure is better than the rest. Apart from this, Morocco and Egypt are also have better infrastructures. However it is worthy to note that many countries of the world are doing well in doing better technological systems in such African countries. Here Fig. 13.5 provided the top countries in respect of most bandwidth of the past (1986, 2001 and 2014) (Hilbert 2016).

And the data shows that there is a significant growth in percentage of bandwidth enhancement in rest past of the world than what happened in 2000 and the same also noted in the developing part also.

## 13.5 ICT in Agriculture: Developing Countries Context

According to most of the study it is noted that there is problem in adequate and proper internet services. ICT and Computing skill is important, but it is noted that in the rural and urban areas, this 'skill' is also an important issue.

### *13.5.1 Inadequate Personnel and IT Infrastructure*

Issues related to the inadequate accessibility of ICT among the rural farmers, computing literacy, internet literacy also valuable concern in this context of agriculture supported by the ICT and Computing. Inability of government services and effort in the developing countries may also consider as valuable in developing Agro Informatics practice in real sense. As far as infrastructure is concerned among aspects, few important are as follows:

- Access to land,
- Average extension services,
- Minimum supportive policies,
- Less agricultural technological utilizations,
- Less amount and proper training and education.

It is an important fact that ICT is an effective solution to the farmers and agriculture industry. There are issues in this context viz. marketing linkages, poor information management, and involvement of the women in ICT in the field of agriculture (Goraya and Kaur 2015; Milovanović 2014; Othman and Shazali 2012).

Proper telecommunications are not only issue in rural areas but also issue in urban areas in most developing countries. Therefore there is not only gap in existing technology but also gap in new technology. According to few expert, due to unsuitable ICT policies, urban and rural communities are unable to get the benefits of the ICT initiatives. It was further argued that poor implementation in support of ICT also reflected to the rural farmers. Further increase of sophisticated software with enhanced human capital requirements also consider as important in Agro Informatics practice.

Inadequate as well poor and unstable power supply, cost of IT infrastructure, awareness, time, cost of other technologies including integration of the system also consider as major constraints of ICT in Agriculture. Furthermore the design of MIS is also needed and in developing countries even such strategies are not proper. Mobile phones consider as important and valuable for enhancing communication and in this regard the developing countries are really important issue. Bad mechanisms and infrastructure is always reason for hampering the agricultural knowledge and led the data redundancy and duplication. Therefore in many developing countries this is an issue and further it can lead a critical role in hampering ICT in governance and agricultural growth; and researchers have reported in Southern Ethiopia that this is happened strongly. The internet users also considered as important as discussed previously. Here more is depicted in Fig. 13.6. It shows the increasing number of internet users per 100 inhabitants (Source ITU report and Wikipedia).

### 13.5.2 Power Supply, ICT Skills, Communication Skills

Implementation of ICTs in Agricultural Sector in Tanzania also similar; however here apart from the Information Technological support another important consideration is power and electric supply and systems (Adão et al. 2017; Paul 2013). This is also common in other developing countries in other parts of the world as well even in India also. Experts also expressed about the lack of knowledge of various aspects of agriculture viz. current happening, way to get information, technologies and machines available for the agriculture and similar sectors. Further lack of knowledge on different fields and language skills are also issue in some of the developing countries; as the farmers are less aware about the English and other international languages in which most of the agricultural contents are published. In developing countries another issue is non inclusion of the Agro IT in the curricula of Agriculture and allied fields and even backdated curriculum in the agriculture and allied fields. The researchers expressed that Information Technology could be helpful in agricultural development and indirectly the life in rural areas as well. Transaction costs, transport costs, marketing cost in some of context in the developing countries are

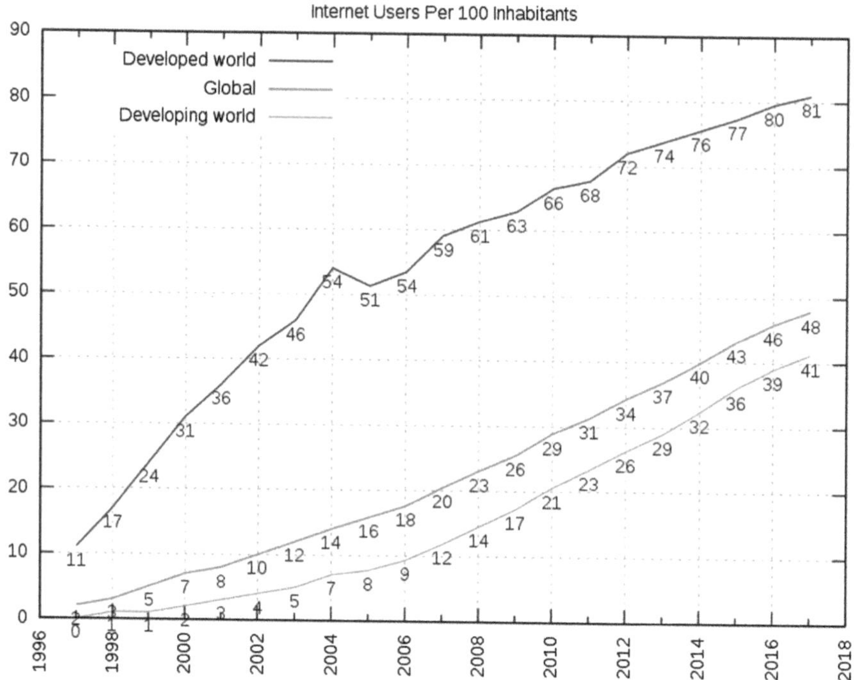


**Fig. 13.6** Internet users per inhabitants

higher which ultimately effects in proper Agro ICT development. And this is many ways due to lack of uses of IT in agriculture. Farmers always need to get information on seeds, fertilizers, pesticides etc. and here ITC is an important tool.

According to study it is noted that in Africa mobile and web services based call centres have been established in some countries to improve access to extension services. In India also a project involves with the agriculture is rising i.e. e-Choupal for the farmers with the help of dedicated Internet kiosks. With this the sharing of best practices in agriculture become easy which leads the improvement in the productivity, and financial benefits and so on.

The digital technologies become easy in communication and information sharing and for such it the need of hour to introduce ICT in agricultural systems.

### *13.5.3 Improving Market Access and Financial Aspects*

ICT is helpful in knowing information on markets and business related affairs. Prices of the agro products, about the consumers, and traders, markets etc. can become important using ICT. They also reflect changing consumption patterns and contain information that can be used by farmers when they decide what and how much to produce. With increased access to mobile phones, farmers can better plan production and investments, based on supply-and-demand fundamentals, thus increasing market efficiency. For example, the *Reuter's RML Information Services* was started in India at Maharashtra in 2007 with SMS based updates on prices, markets, and weather etc. to the farmers. The Smartphone based application was also been offered to the 1,300 markets across India.

e-Choupal in India offers various financial services like matching services, commodity exchanges, virtual trading, assess quality. Further it is only provide the transparency and efficiency in agro based job and market scenario. The researchers stated that in Ghana and Uganda the ICT based market price information is available by the payment to cover the cost of the service. Researchers however did not find any further promotion of the farmers by the apps.

*Esoko* is another web based services and advisory call centres and consider as an important initiative of Grameen Foundation in Uganda, this initiative overcomes various challenges in information transfer regarding the agricultural affairs to reach rural with access and affordability of ICTs. Community Knowledge Worker here doing a lot to enhance the technology toward poor communities.

*Digital Green*, An International Development Organization is also offered the innovative digital platform to improve lives of rural communities in South Asia and Sub-Saharan Africa using ICT based services. The Digital Green is associated with the local public, private and civil society organizations to offer agro services to improved agricultural practices, livelihoods, health etc.

Regarding the transfers and payments, credit, savings, financial services etc. also ICT are very much effective and useful. ICTs is helpful in developing rural communities the financial services apart from the unconventional methods. And in this regard *M-PESA is an important example.* With this the farmers of many developing countries (viz. Kenyans, Nigerians, Tanzanian etc.) are able in sending money; therefore such ICT based financial services helps in agricultural systems indirectly. *DrumNet* in Kenya is another example with ICT platforms which helps the farmers in banking and microfinance institutions activities and so on. *Kilimo Salama is another example which* allows Kenyan smallholder farmers regarding the technological benefits viz. smartphones etc. which can indirectly helpful in developing agricultural systems (Channe et al. 2015; Na and Isaac 2016).

### *13.5.4 Climate and Early Warning Systems and Developing Countries*

Climate and weather always impacting agriculture, and indirectly it is also deals the aspects of the food security. Proper and prior knowledge on weather and agricultural system can lead the successful or failed harvesting system and here ICT-based tools can be effective especially the early warning systems. Timely updates on local meteorological conditions are the need of hour in developing countries as well. There are various challenges and issues in this regard. A suitable example is *AfricaAdapt* in Senegal which deals in information on various aspects viz. climate change adaptation from researchers, policymakers, and it will be helpful in community practice of agro systems.

One of the prime initial example of such weather forecasting systems in regard to agriculture may be World Food Studies (WOFOST) simulation model and this is developed by the Centre for World Food Studies (CFWS) in association with the University of Wageningen. This is also able in calculate production levels including soil and weather conditions.

ICT-mediated early warning and disaster information systems as a project also came into force as FAO initiated *Avian Influenza alert system* and practiced in the Bangladesh. Here mobile technology being used to find out the deadly avian (H5N1) virus. FAO and Google have also been initiated for gather geospatial data and analysis through *Open Foris*. Here Google's Earth Engine also used with the coverage of forestry, land cover and land use. Here the open source app allows smartphone to monitor a piece of land including assessing of deforestation and forest degradation. These services are offered in developing countries also but due to the shortage and lack of knowledge of such residents it may be difficult to pursue the program with large scale.

Foodborne illnesses pose a serious health threat and here proper risk-management tool are required and it is further to be noted that to find out the animal disease including monitoring and controlling effectively various modern electronic tools been available and initiated. In Agro Marketing also ICT based services are important viz. *TraceNet* which is able in the certification for export of organic products from India is also important.

## 13.6 ICT in Agriculture in Developed Countries in Respect Developing Countries

As far as digital divide is concerned it is very important to note that the scenario is different in developing and developed countries; and obviously in agriculture also this is not an exemption. There are various challenges in implementation of the digital technologies in agriculture and ultimately it affects the scenario of developed economies and the developing world. In the field of agriculture we can noted with various famers and agro related activities and with large, medium and small input suppliers, traders, processors and retailers. Here ICT throughout in various areas are applicable and increasing.

Farmers are using various emerging IT and Computing tools in diverse areas viz. soil analysis, irrigation, weather forecasting, and so on. In the context of technological development the same can be noted in the field of agriculture as well.

In the developing countries there are certain developments in the field of agriculture and Precision Agriculture can be consider as important in this regard. There are certain issues in environmental impacts from the perspective of farming. This technique and smart farming is applicable in Central and Northern Europe, in many developed Asian Countries, in Americas mainly USA, Australia, New Zealand. Here the uses of Controlled Traffic Farming may be considered as important in crop damage and soil compaction using GNSS technology and decision support systems. In the Australia, UK also various strategies are resulting less input costs and increase crop yields.

Precision Livestock Farming is also associated with the ICT and Computing. In automatic monitoring of the animals including milk and egg production find out their physical environment also ICT is important in many parts of the world doing with viz. England, New Zealand. Proper ICT uses and further need for research may be consider important to increase the amount of Agro Production. In the European Union European research programs are engaged with the in ICTs applications in large scale and even in smaller farms. Precision Agriculture in recent past has also associate with the emerging technologies viz. Cloud Computing, Big Data etc. for enhancing the capacity. Big Data as deals with the large and complex data sets and here Agricultural areas therefore in real and in future requirement of Agriculture Big Data is worthy. In previous and future predictions Big Data using Machine Learning is useful. In efficient decision making various tools are useful which needed in agricultural development with biodiversity protection. In United States two data exchange platforms *FieldScripts* and *Farmers Business Network* (FBN) are considered as important (Gómez-Chabla et al. 2019; Paul et al. 2015).

The e-Agriculture Community of Practice founded in 2007 by the FAO to solve the problem in digital divide in the field of agriculture and promoting healthy ICT practice. Furthermore, the recommendations of WSIS (World Summit on the Information Society), of United Nations are also considered as important by the FAO.

FAO is engaged with the development and facilitation of ICT in the field of agriculture, healthy global community of practice, exchange of information, ideas, and resources related to the Agricultural Informatics. It is also engaged in rural development via e-Agriculture Community development by helping decision making using ICT. In developing sustainable agriculture, food security also FAO's initiative by real Agro Informatics practice is noticeable. Various information and communication specialists, academic and professional researchers, farmers, business and policy people, Agro ICT development consultant are associated in e-Agriculture initiative of FAO which is helping different countries though the developed countries are getting well benefits due to their already having basic infrastructure. focuses In e-Agriculture Community Practice of FAO various other establishments also engaged in solid sense for healthy practices such as

- Various UN Agencies
- Governments of various countries
- Universities across the globe with Agro ICT interest and development
- Research organizations and institutions with Agro ICT interest and development
- NGOs with Agro ICT interest and development
- Agro and Agro IT based farmers' organizations

- General and wider community and association.

The e-Agriculture Community of Practice, with FAO, the ITU (International Telecommunication Union), and the CTA initiated 'National E-agriculture Strategy Development' to develop national e-agriculture strategies in Asia–Pacific and Central Asia regions. And by this many countries got benefited but mostly are developed countries and few are developing due to their lack of ICT infrastructure and initiatives and other issues.

FAO and ITU are also engaged with the *FAO-ITU e-Agriculture national capacity building efforts* regarding the implementing national e-Agriculture strategies to overcome challenges and in agriculture sector goals. Here the e-Agriculture Strategy Guide of FAO and ITU can be consider as important guidelines.

Here multi-stakeholder approach and involvement of the national stakeholders of the countries and also from the government agencies, private and technological companies, Internet and telecom providers, civil society organizations etc. can be consider as important and vital. FAO and ITU is even also did in developing ICT in Agriculture in many developing countries such as Sri Lanka, Bhutan such as in action plan designing and execution, monitoring and evaluation mechanisms. In the countries like Fiji, Philippines, Papua New Guinea and Vanuatu also the joint initiative of FAO and ITU are noticeable.

The WorldBank Project as 'The ICT in Agriculture Sourcebook' may be consider as another on-line practical guide in agro ICT practice, finding the current trends, appropriate interventions. It offers empirical knowledge on agriculture of ICT for the farmers, practitioners, decision-makers. This is employed by many developed countries and gained a lot. Furthermore in 14 sub-sectors of agriculture this The ICT in Agriculture Sourcebook' can be considered as important viz.

- Rural finance.
- Markets.
- Agribusiness value chains.
- Extension.
- Innovation systems.
- Farmers' organizations.
- Agricultural marketing.
- Agricultural risk management.
- Food safety and traceability.
- Land administration and management.

The ICT in Agriculture Sourcebook' is consist with 200 project-based case studies which can be helpful in developing ICTs on agricultural development and partially in enhancing economic development, in minimizing rural poverty reduction, long term development. Here various complex set of policy, investment, capacity-building aspects, sustainable ICT infrastructure development, services for the rural economy related affairs been noted.

InfoDev, is another World Bank associated and global trust fund initiative on the Development of entrepreneurship in Agricultural sectors and it has70+ countries involvement. Through this innovative pilot program is intended for the financing, business training on ICT uses in Agriculture with reference to the climate technologies, digital innovation, promotion of the jobs and services in agricultural sectors.

Various emerging technologies are considered as important in the InfoDev activities for better agro informatics practice and also in strengthening social inclusion. The Digital Entrepreneurship integrated with the mobile application enhancing many developed countries in the field of agriculture.

Mobile Application Labs (mLabs)—is another initiative and innovation hubs for digital entrepreneurs. Mobile Social Networking Hubs (mHubs) is also been offered in many countries regarding the offering of training programs in agro ICT, testing facilities, health and financial inclusion, agricultural management, information technology.

mLabs and mHubs, InfoDev etc. are emerging and enhancing the ICT systems in agriculture and many developed and developing countries received the benefits from the same. *Mfarm* can be consider as another example that enhanced the market access in Kenya whereas the *GreenHouse Pro* is another initiative that facilitate productivity of greenhouse farming. Another app and systems *MkulimaBima* is an example of insurance for the farmers.

InfoDev's is surrounded by IT and supporting technology to development of a digital economy. And this can led the digital technology clusters and emerged in different cities of various countries viz. Bangalore, Berlin, Hangzhou, London, Nairobi, and New York and even in many countries of Africa as well. U.S. Agency for International Development (USAID) supported another project is Fostering Agriculture Competitiveness Employing Information and Communication Technologies (FACET) which is to improve the ICTs for the development of the competitiveness and productivity across sub-Saharan Africa. FACET is enhanced to build collaborative relationships, sustainable, scalable approaches, technical assistance using ICT applications toward agriculture. The e-Agriculture Community of Practice and FACET offers agricultural extension services regarding Agricultural ICT application using videos, webinars, etc. (Aubert et al. 2012; Zamora-Izquierdo et al. 2019).

The Technical Centre for Agricultural and Rural Cooperation is a joint international institutional engagement of the African, Caribbean and Pacific Selected States and also with the European Union (EU) toward healthy ICT strategies, solving ICT policy issues, implementation of ICTs in agriculture strategic implementation. In India, IFPRI (International Food Policy Research Institute) in collaboration with the Centre of Tamil Nadu Agricultural University has been developed the Advanced Agricultural Practice Knowledge Portal for in 2012 as a gateway for knowledge in agriculture and helpful for the farmers, research scientists, agricultural associations, NGO and agricultural entrepreneurs and so on. As a result it is beneficial for the small and marginal farmers and various intermediary organizations.

European Commission's ERA-NET scheme funded ICT-AGRI is designated to develop, strengthen the initiatives to coordinate regional, national and European programmes in agro and allied fields and ran until 2014. It helps in developing Precision Agriculture using ICTs and robotics in agriculture specially in fertilizer, pesticides and water. Further it was also designated in Controlled Traffic Farming, precision livestock farming safety and traceability of food on big and smaller farms. Big Data is managing about the complex system including technologies of various types and in agriculture Big Data applications have raised in many developed countries for healthy inclusion. Open Foris is a system specific geospatial data through Google Earth. Here Bing Maps and Google Earth Engine also used in satellite imagery in diverse agricultural sector viz. land and forestry assessments, monitoring urban areas etc. M-PESA is another joint initiative of Vodafone and USAID toward better mobile services in the agricultural sector in Kenya, Tanzania and Mozambique. It has used remote crowdsourced data-collection method and important in crops specialization in producing. Here structured and spatial data are useful in farmers and their products.

Therefore various projects have been completed in respect of ICT for the Development; and various are running worldwide in response to developing Agricultural Information Technology Practice. Many projects are global but in certain cases only the developed countries are able to get wider benefits for the Agro ICT development (Adetunji and Joseph 2018; Paul et al. 2020).

## 13.7 Conclusion

Agricultural Information Systems and IT practice are rising gradually throughout the world. There are certain technologies being used rapidly by the farmers. It is worthy to note that it is difficult to employ all kind of technologies and systems which are costly and difficult to implement by all the countries and agricultural bodies. The rising applications of the IT and Computing also effected agricultural sectors. Basic tools, technologies, and systems are in generally used in the Agro Informatics also viz. basic mobile phones uses to get information about the agricultural crops, products, and market. Since Robotics, HCI, Cloud Based Systems, and Drones may not be suitable for all the counties, agro associations, and farmers, in such context basic free mobile based Agro Information Services can be an important alternative. Furthermore there are huge differences in existing infrastructure, ICT Systems, financial conditions, skill sets among the farmers and associated with the agriculture. A proper policy, framework, etc. are most suitable and need of the hour in complete and holistic development of the Agricultural Informatics practice and toward a healthy ICT.

## References

Abbasi AZ, Islam N, Shaikh ZA (2014) A review of wireless sensors and networks' applications in agriculture. Comput Stand Interfaces 36(2):263–270

Adão T, Hruška J, Pádua L, Bessa J, Peres E, Morais R, Sousa JJ (2017) Hyperspectral imaging: a review on UAV-based sensors, data processing and applications for agriculture and forestry. Remote Sens 9(11):1110

Adetunji KE, Joseph MK (2018) Development of a cloud-based monitoring system using 4duino: applications in agriculture. In: 2018 international conference on advances in big data, computing and data communication systems (icABCD). IEEE, pp 4849–4854

Ahmad T, Ahmad S, Jamshed M (2015) A knowledge based Indian agriculture: With cloud ERP arrangement. In: 2015 international conference on green computing and Internet of Things (ICGCIoT). IEEE, pp 333–340

Aubert BA, Schroeder A, Grimaudo J (2012) IT as enabler of sustainable farming: an empirical analysis of farmers' adoption decision of precision agriculture technology. Decis Support Syst 54(1):510–520

Babu SM, Lakshmi AJ, Rao BT (2015) A study on cloud based Internet of Things: CloudIoT. In: 2015 global conference on communication technologies (GCCT). IEEE, pp 60–65

Balamurugan S, Divyabharathi N, Jayashruthi K, Bowiya M, Shermy RP, Shanker R (2016) Internet of agriculture: applying IoT to improve food and farming technology. Int Res J Eng Technol (IRJET) 3(10):713–719

Bauckhage C, Kersting K (2013) Data mining and pattern recognition in agriculture. KI-Künstliche Intelligenz 27(4):313–324

Channe H, Kothari S, Kadam D (2015) Multidisciplinary model for smart agriculture using internet-of-things (IoT), sensors, cloud-computing, mobile-computing & big-data analysis. Int J Comput Technol Appl 6(3):374–382

Gill SS, Chana I, Buyya R (2017) IoT based agriculture as a cloud and big data service: the beginning of digital India. J Organ End User Comput (JOEUC) 29(4):1–23

Gómez-Chabla R, Real-Avilés K, Morán C, Grijalva P, Recalde T (2019) IoT applications in agriculture: a systematic literature review. In: 2nd international conference on ICTs in agronomy and environment. Springer, Cham, pp 68–76

Goraya MS, Kaur H (2015) Cloud computing in agriculture. HCTL Open Inte J Technol Innov Res (IJTIR) 16:2321–1814

Guardo E, Di Stefano A, La Corte A, Sapienza M, Scatà M (2018) A fog computing-based iot framework for precision agriculture. J Internet Technol 19(5):1401–1411

Hilbert M (2016) The bad news is that the digital access divide is here to stay: domestically installed bandwidths among 172 countries for 1986–2014. Telecommun Policy 40(6):567–581

Holster HC et al (2012) Current situation on data exchange in agriculture in the EU27 & Switzerland. agriXchange, pp 1–15

Kamble SS, Gunasekaran A, Gawankar SA (2020) Achieving sustainable performance in a data-driven agriculture supply chain: a review for research and applications. Int J Prod Econ 219:179–194

Kajol R, Akshay KK (2018) Automated agricultural field analysis and monitoring system using IOT. Int J Inf Eng Electron Bus 11(2):17

Khattab A, Abdelgawad A, Yelmarthi K (2016) Design and implementation of a cloud-based IoT scheme for precision agriculture. In: 2016 28th international conference on microelectronics (ICM). IEEE, pp 201–204

Liu S, Guo L, Webb H, Ya X, Chang X (2019) Internet of Things monitoring system of modern eco-agriculture based on cloud computing. IEEE Access 7:37050–37058

Manos B, Polman N, Viaggi D (2011) Agricultural and environmental informatics, governance and management: Emerging research applications. In: Andreopoulou Z (ed) IGI global (701 E. Chocolate Avenue, Hershey, Pennsylvania, 17033, USA)

Milovanović S (2014) The role and potential of information technology in agricultural improvement. Econ Agricul 61(297-2016-3583):471–485

Muangprathub J, Boonnam N, Kajornkasirat S, Lekbangpong N, Wanichsombat A, Nillaor P (2019) IoT and agriculture data analysis for smart farm. Comput Electron Agric 156:467–474

Na A, Isaac W (2016) Developing a human-centric agricultural model in the IoT environment. In: 2016 international conference on Internet of Things and applications (IOTA). IEEE, pp 292–297

Nandyala CS, Kim HK (2016) Green IoT agriculture and healthcare application (GAHA). Int J Smart Home 10(4):289–300

Nayyar A, Puri V (2016) Smart farming: IoT based smart sensors agriculture stick for live temperature and moisture monitoring using Arduino, cloud computing & solar technology. In: Proceedings of the international conference on communication and computing systems (ICCCS-2016), pp 9781315364094-121

Ojha T, Misra S, Raghuwanshi NS (2015) Wireless sensor networks for agriculture: the state-of-the-art in practice and future challenges. Comput Electron Agric 118:66–84

Othman MF, Shazali K (2012) Wireless sensor network applications: a study in environment monitoring system. Procedia Eng 41:1204–1210

Ozdogan B, Gacar A, Aktas H (2017) Digital agriculture practices in the context of agriculture 4.0. J Econ Financ Acc 4(2):186–193

Pau1 PKMG, Chaterjee D (2014) Information Systems & Networks (ISN): emphasizing agricultural information networks with a case study of AGRIS. Sch J Agric Vet Sci 1(1):38–41

Paul PK (2013) Information and knowledge requirement for farming and agriculture domain. Int J Soft Comput Bio Inf 4(2):80–84

Paul PK et al (2015) Agricultural problems in India requiring solution through agricultural information systems: problems and prospects in developing countries. Int J Inf Sci Comput 2(1):33–40

Paul PK et al (2016) Cloud computing and virtualization in agricultural space: a knowledge survey. Palgo J Agric 4(2):202–206

Paul PK et al (2015) Information and communication technology and information: their role in tea cultivation and marketing in the context of developing countries—a theoretical approach. Curr Trends Biotechnol Chem Res 5(2):155–161

Paul PK, Sinha RR, Pappachan Baby KS, Shivraj BA, Mewada S (2020) Usability engineering human computer interaction and allied sciences: with reference to its uses and potentialities in agricultural sectors: a scientific report. Sci Rev 6(7):71–78

Rezník T, Charvát K, Lukas V, Charvát Jr K, Horáková Š, Kepka M (2015) Open data model for (precision) agriculture applications and agricultural pollution monitoring. In: EnviroInfo and ICT for sustainability 2015. Atlantis Press

Teye F, Holster H, Pesonen L, Horakova S (2012) Current situation on data exchange in agriculture in EU27 and Switzerland, ICT for agriculture, rural development and environment. In: Mildorf T, Charvat Jr C (eds) Czech Centre for Science and Society Wirelessinfo, Prague, pp 37–47

TongKe F (2013) Smart agriculture based on cloud computing and IOT. J Converg Inf Technol 8(2):210–216

Tsekouropoulos G, Andreopoulou Z, Koliouska C, Koutroumanidis T, Batzios C (2013) Internet functions in marketing: multicriteria ranking of agricultural SMEs websites in Greece. Agrárinformatika/j Agric Inf 4(2):22–36

Zamora-Izquierdo MA, Santa J, Martínez JA, Martínez V, Skarmeta AF (2019) Smart farming IoT platform based on edge and cloud computing. Biosys Eng 177:4–17